Kick the Hay Habit
A practical guide to year-around grazing

Kick the Hay Habit:

A practical guide to year-around grazing
by
Jim Gerrish

A division of Mississippi Valley Publishing Corp.
Ridgeland, Mississippi

First printing January 2010
Second printing June 2010
Third printing August 2011
Fourth printing March 2013

Library of Congress Cataloging-in-Publication Data

Gerrish, Jim, 1956-
    Kick the hay habit : a practical guide to year-around grazing / by Jim
        Gerrish. -- 1st ed.
    Includes index.
    ISBN 0-9721597-4-6 (alk. paper)
    1. Grazing--United States. I. Title.
    SF85.3.G47 2010
    636.08'45--dc22
                        2009054118

Cover design by Steve Erickson, Ridgeland, Mississippi
Cover photo by Jim Gerrish
Graphics by Jim Gerrish

Manufactured in the United States.
This book is printed on recycled paper.

Table of contents

This book is dedicated to my wife and children without whose help I would not have gotten nearly so much done in my time here on earth.

Dawn Ann (Larson) Gerrish
Chelsea October Gerrish
Ian Jackson Gerrish
Galen James Gerrish
Cavan Hamilton Gerrish

Chapter 1
The culture of hay

"Make hay while the sun shines."
"It was a romp in the hay."
"It's time to hit the hay."
"Let's have a hay ride."

You don't have to be a farmer or rancher to understand these common expressions from everyday life. Our city cousins are just as likely to use these expressions as we are. Hay, with all of its connotations, is an ingrained part of our culture. Ingrained to the point that most farmers and ranchers don't believe they can live without it.

And why shouldn't it be a part of us? Humans have reportedly been drying forage during the growing season to feed in the dormant season for at least 4000 years. As long as man has kept domesticated livestock in the climates of northern Europe and Asia, he has been making hay. Roman cavalry horses ate alfalfa hay in Gaul in the first century AD. The first depiction of hay in art was on the emperor Trajan's memorial column in Rome. Trajan led his legions into Dacia, modern Romania, and found haymaking to be a long established practice among the Dacians who fed it to sheep, cattle, and horses.

Paintings by the Great Masters of the Renaissance show stacks of hay in their scenes of daily life in country villages throughout Europe. When I was in college, I had a print of

Pieter Bruegel's The Hay Harvest on my wall. It is a painting of an idyllic countryside with seemingly well fed and well dressed peasants happily going about the harvest. Very reminiscent of the myopic view of many of today's farmers and ranchers.

From early Roman times through the early 18th century, hay making changed very little. Cutting hay was hard manual labor carried out with a sickle. A sickle requires bending low to the ground and after a day of hard toil perhaps a third to half of an acre was lying on the ground. Then in the early 1700s, hay cutting was revolutionized by the invention of the scythe. A strong man on the scythe could lay down swath after swath all day long and, at the end of the day, over an acre of the tall, swaying grass would be on the ground.

After men had cut the hay with sickle or scythe and it had cured for a few days, it was time for the women and children to come with large wooden rakes and gather it into piles. Then it was the men's turn again as they forked it from the raked piles with large wood-tined pitch forks onto carts and wagons. It was a labor that lasted through most of the summer for farmers in northern Europe and British Isles and Ireland. It was a rite of passage for every farm boy. The day he took the pitchfork to the hay meadow instead of a rake was the day he could begin to call himself a man. When he spent a full day swinging a scythe was the day he could have a pint at the local pub along with the other men.

I'm sorry, ladies, there was no gender equality in those days.

As soon as the Industrial Revolution touched Europe and America, inventors sought ways to mechanize the harvesting process. The scythe was replaced by the horse-drawn sickle bar mower in the mid 1800s. Although often credited to Cyrus McCormick, the first reciprocating mower was invented by Hiram Moore in 1841. Stationary wire-tie balers began to displace the beehive-like hay stacks by the late 19th century. Throughout the 20th and into the 21st century, the agricultural-industrial complex has developed more and more elaborate equipment and processes to satisfy our lust for making hay. Technology had successfully taken a brutal, labor demanding task and turned it into a simple one-person job.

I grew up making hay. I still love the rich fragrance of fresh mown hay on a summer's evening. I can still enjoy watching the tall grass fall behind a sickle-bar mower. Even as I know my youth has passed, I still revel in the invigoration from stacking small square bales. Though, now I tire after the first 100 bales of the day instead of the first 1000 bales.

Some of my best memories from my teenage years are working on hay crews around our county in southern Illinois. From pranks we played on each other and the farm owner, to the late afternoon lunches the German farm wives brought out to the field. Long summer days that started before sunrise and ended when the last wagons were unloaded by the naked light bulb hanging from the barn's ridgepole.

Part of the culture of hay was neighbors working together to get the hay harvested and stored between rain showers. For me it was our four-man crew picking up and stacking 1500 bales a day and not a one of us over sixteen. We were just proud to be doing a man's work for a buck-and-a-half an hour. If we could get a littler kid to drive the tractor for us, the four of us could all work and not have one of us doing the sissy job of driving. Better yet we'd get the farmer's daughter to drive the wagon.

It was seeing the car coming up the dusty field lane precisely at 4:30 in the afternoon laden with sandwiches, cookies, and lemonade. If there was a shade tree nearby we would go there, but more often it was just dragging a few bales into the shade of the stacked wagon and sitting down for a snack. If you ate three or four sandwiches, the more the ladies liked you and wanted you to come back to work for them again. If you only ate one, the boss would think you must have been slacking off and he wouldn't want you back.

Then there was the day we were working for a new family for the first time. They had a good German name and the wife showed up at 4:30 right on time. But when we unwrapped the sandwiches, they were mayonaise and banana sandwiches and she brought nothing to drink. What kind of a cruel joke was that? To bring four or five hungry boys such tripe in the middle of a 100 degree afternoon? Nobody could eat the darn things and we went back to work, vowing to get even.

This man had a nice big barn with a good solid loft floor but he was making us stuff bales into the chicken house, smoke house, and whatever other odd shed was around. Typically the doors were narrow and you could only put a bale in endwise. Everywhere we could, we finished the sheds off with the cut side of the bale to the door, extending past either side, and the last bale stuffed in at the top.

When the old man opened up the door and looked, there was nothing in front of him but a solid wall of hay with no twines visible and unable to push the top bale up out of the way. He looked at the wall of hay and said, "%$*&@, you boys sure filled up my *&%$#@& barn." Although that was over 30 years ago I can still hear him to this day. Next year when we went back there, he let us stack hay in the loft and his wife brought us good bologna and salami sandwiches. And the story became part of our personal hay culture.

My dad bought one of the very first large round balers in our part of the state and we began custom hay making in a big way. My first four years of college were paid by hay making. One summer I put 6700 bales through a Vermeer C-model baler pulled by a cabless tractor and over half the hay was red clover. Dusty and nasty stuff to bale up, but by and large, I loved it. I think that baler cost us $4200 in 1973 and we charged $5 or $6 per bale, depending on how far we had to travel. We also ran two little square balers, one twine tie and one wire tie. The same year I did 6700 round bales, we did over 100,000 little bales. I definitely came out of a hay culture.

When I went to work at the University of Missouri - Forage Systems Research Center in 1981, our main social event of the season was Hay Day. It was an extravaganza of new paint and black smoke. A day when area farmers could come and see the latest and greatest in hay making equipment. Usually several hundred people would be there and several people always went home with a new toy. We quit doing Hay Day in the early 1990s when it became apparent to Dr. Fred Martz, the station superintendent, and myself that we could no longer in good conscience encourage farmers and ranchers to buy hay equipment. Our culture was beginning to change.

Hay and haymaking have been an integral part of animal agriculture for millennia, so why are so many livestock producers these days trying to do away with making and feeding hay? Quite simply, because the cost of hay has grown to exceed its value.

Why don't you install a stock water distribution system? Too busy baling hay.

Why don't you install subdivision fencing in your pastures? Too busy baling hay.

Why don't you move cattle to a new pasture every day? Too busy baling hay.

Why don't you go to a grazing school? Too busy baling hay.

And the litany of the hay addict goes on and on.

There must be a better way.

It's time we kicked the habit. The Hay Habit.

Chapter 2
The high cost of making hay

Before the industrial revolution, making hay was entirely a hand labor process. A simple scythe was all that was needed to mow the hay. A wooden rake was used to gather it after it had dried. A wooden pitchfork was needed to pile it into stacks or to load it on a cart or sledge to be piled near the hut or barn to be fed later. The needed labor came from a large family. The costs of labor were the shelter and food you provided for your children. Then we learned family planning. The cheap family labor was gone and we began to hire our labor.

Usually the hay field was a natural meadow lying in the lowlands on a deep, naturally fertile field. Winter snow melt and spring rains often caused flooding of the meadows, resulting in deposition of nutrient rich silt and organic debris on the fields. This was nature's cheap fertilizer program. As our herds grew and we needed more hay, we left the riparian meadows for the upland pastures and began our depletion of the soil. Over the passing centuries we learned the value of manure, wood ash, and lime to rejuvenate the soils. And eventually we learned how to make synthetic fertilizers and then we bought our soil fertility from the co-op.

First man learned to harness horses and then he learned how to harness horsepower. We came to rely on the machines pulled by horse and horsepower and we gladly gave up the hand labor of hay making. With fewer children and labor-saving machines, we fell in love with the mechanization of hay making. Now we could do in a day what had taken the whole family weeks to accomplish and we bought bigger and better machines to harvest our hay.

The Industrial Revolution called more and more people to work in the factories and shops of the cities. Many rural folk left the farms for the perceived easier and more comfortable life in the cities and the promise of a weekly paycheck. As agriculture also industrialized, fewer and fewer farmers were needed. Factory wages were often low, but the governments knew hungry people were not happy people, so the cost of food must be kept low and the factory workers would remain content. To protect industry we had farm programs to subsidize production so food would always be cheap.

Americans became accustomed to spending less and less of their income for essential food. If American farmers could not produce it cheaply enough, we could always import it from some country with lower wages or less stringent regulations. Thus, we grew into a world economy where we must all be competitive with the lowest cost producer.

Then one day we woke to find what had worked in past years or centuries no longer worked in the 21st century. The cost of making hay had exceeded its value to our business. Yet the vast majority of American farmers and ranchers continue to make hay today. We can come up with a number of reasons why we continue to make hay even though it is a money-losing process.

One reason might be they are not profit motivated producers. While this may sound shocking to most commercial livestock producers, many surveys conducted around the USA have found only about half the beef cattle producers in this country are profit motivated. Half raise cattle for other reasons including personal enjoyment and lifestyle, an alternative to mowing their country estate, or something for their kids to do to learn about responsibility. With motivations like this, some hay will continue to get made no matter the cost.

Another reason some continue to make hay is because that is what they have always done and they cannot conceive of raising livestock without hay. They might be profit motivated, but fear of change and the need to maintain their comfort zone requires some other alternative to kicking the hay habit. The usual solution is for someone in the family to get an off-farm job so they can afford to either make or buy hay.

I think the main reason most producers continue to make hay is they simply have no idea what it is really costing them. Most producers don't know what it costs them to produce a pound of beef or lamb or a gallon of milk, let alone what individual inputs cost. Until you know what your costs are, you cannot manage them. Think of all the ingredients needed for making hay and then think of the costs associated with each ingredient.

So what does it cost to produce a ton of hay?

We must have land to grow the crop. That land may be owned or leased or maybe we buy a standing crop from a neighbor. Any way we do it, at some level there is a land cost. Leased land costs are easy to figure as you probably actually write a check to pay that bill. If the bill is for land that gets used for different enterprises, separate out the hay acres and assign a cost to that enterprise. If you pay $40/acre to rent hay land and it produces four tons/acre, the land cost is $10/ton plus interest. It's really that simple.

If you own the land, is it paid for or are you still making payments? Do you treat land ownership as a separate investment business or does the hay produced have to make the payment? Some people like to assign a percent return on their land investment as the annual land cost. In most areas, land price is unrelated to ag production level so this approach would make almost any conventional ag operation completely unprofitable, unless a very low rate of return is used (e.g. <2%).

The simplest way to assign a land charge to a specific enterprise is using the local prevailing rental rate for that type of land. If the field produces a crop of hay and then is grazed, prorate the rent based on the relative yield as hay or pasture. I prefer using the animal use record for the proration basis. If you harvest a ton of hay and it feeds 60 cows for one day it has a different value than if it fed only 40 cows for a day. Carelessness in feeding hay can lead to excessive waste that makes it even more expensive. The real value is determined by the animal product generated.

We must have forage growing on the land to harvest. Perhaps there is already a stand of native plants and it is very sustainable, thus our cost is very low. Or maybe we have a piece of bare land and we will seed it to some crop for hay harvest. That cost may be less than $50/acre or it may be over $200/acre. The annual cost of the crop depends on how long the seeding survives as a productive field. If it is an annual crop, the establishment cost may be very high, but if it lasts twenty years the cost may be quite low.

The cost of establishing a perennial forage crop is often ignored by farmers and ranchers calculating their hay production cost. If a hay field is torn up and reseeded every five years, as is common with alfalfa in many parts of the country, the cost of establishment is prorated over the life of the stand with interest charged for the same time period. If it cost $150 to seed the new hay field and it is replaced every five years, the establishment cost is $30 annually plus accrued interest. At 8% interest, the annual cost is $37.70/acre or about $9.55/ton using the 4 ton/acre example.

Each ton of hay removes a certain quantity of minerals from the soil. If we transport that hay away from the field where it was produced, those nutrients are gone from that ecosystem. If we are blessed with an alluvial field with naturally high fertility, our cost may be fairly low. But if our field is a sandy soil low in organic matter, the nutrients must be replaced with each crop or productivity will fall. Each ton of hay on average contains 40-50 lb of N, 10-14 lb of P_2O_5, and 40-50 lb of K_2O, plus an array of other minerals.

If the hay is fed somewhere other than on the field

where it was harvested, those nutrients have been removed from the field and must be replaced. Alfalfa and other legume hays do not need N fertilizer, but grasses without companion legumes need some source of N to be productive. In 2008 we saw N fertilizer prices soar to over 70cents/lb in many locations. A four-ton yield of grass hay would require about 160-200 lb-N/acre at a cost of $120-150/acre or $30 to $37/ton. The P & K content value, based on a 4-ton alfalfa yield, in 2008 was about $100-120 or $25-30/ton.

Because we don't use our children with scythes anymore, we must have equipment to harvest the hay and tractors to pull or push it. Swathers and mower conditioners, rakes and tedders, and balers, large or small, square or round. We can buy it new or we can buy it used. We can lease it or borrow it. However we get it, it comes at a cost.

Many small operators don't believe they have significant equipment costs because they bought it all used at local farm sales. Very often their equipment cost per ton of harvested hay is higher than their mega-farming neighbors with all new equipment. Why? Because the low tonnage of hay they produce has a very high equipment cost per ton. Using your IRS Schedule F depreciation schedule is a good way to determine the equipment costs associated with a particular enterprise. Just take it one step farther and assign a percentage of each tractor or piece of equipment to individual enterprises.

Then we need a place to store the hay. In the eastern half of the USA and Canada, the wet climate prevents hay from being stored outdoors as is commonly done in the arid West. One of the purported advantages of the large round bale when it was first introduced was that it could be stored outdoors in high rainfall areas and not spoil. Unfortunately that didn't work out quite as well as planned. So we invented net wrap and bale wrappers and another whole set of tools to combat spoilage. And, once again, it all came at a cost.

Then we need to feed the hay, which means we must have equipment to haul it out again. We need feeders to reduce the amount the animals waste. We invented bale processors and tub grinders to make lousy hay more palatable. Someone has to haul the hay to the livestock so labor costs are added.

And in the final round, just to add insult to injury, we had to haul off the manure, at a cost.

Everyone's cost of producing hay is a little different and is determined by how each of the costs listed above fits their individual situation. Just for fun, here is a sample budget. You should consider each of the inputs and determine your cost.

These are the clear cut financial costs of making hay. There are two other very significant costs of making and feeding hay that we hardly ever consider. The opportunity costs of your time and your land resources.

HAY PRODUCTION COSTS:								
Acres of hay		200	acres					
Expected yield:		5	tons/acre					
Number of harvests		3						
Weight of bales		1200	lb/bale					
Total hay produced		1000	tons					
Bales / acre		8.3						
								Cost/acre
Land rental or standard fee								$ 65.00
Establishment								$ 60.08
Maintenance fertilizer								$ 196.50
		COST/LB	REMOVAL	APPLY?				
	N	$ 0.55	250	0	$ -			
	P	$ 0.45	60	1	$ 27.00			
	K	$ 0.70	225	1	$ 157.50			
Spreading cost		$ 6.00		2	$ 12.00			
Mowing (cost/operation)		$ 16.00	/A					$ 48.00
Raking "		$ 5.00	/A					$ 15.00
Small square bale		$ 0.60	/bale	0	$ -			
Small square bale handling		$ 0.60	/bale	0	$ -			
Large round bale		$ 9.00	/bale	1	$ 75.00			
Large round bale handling		$ 2.00	/bale	1	$ 16.67			
Large square bale		$ 12.00	/bale	0	$ -			
Large square bale handling		$ 2.00	/bale	0	$ -			
Equipment depreciation			/acre					$ 90.39
Irrigation cost			/acre					$ 40.00
			Cost/acre					$ 606.64
			Cost/Ton					$ 121.33
			Cost/Bale					$ 72.80
			Cost/lb					$ 0.061

Think about the hours, days, or weeks spent harvesting hay in the summer months. What else could you have been doing with that time if it hadn't been devoted to hay making? One of the most ironic comments regarding time and hay making has to do with the many grazing schools or field days I have been involved with over the last 20 years. I cannot count the times a farmer or rancher has told me over the winter and through the spring they plan to come to a particular grazing school and then they don't show up. When I see them later, their excuse is, "I was in the middle of haymaking."

Grazing schools aside, if you weren't making hay, could you do a better job of monitoring your pastures and moving livestock to new paddocks in more timely manner? Could you find the time to build the fences and install the water systems you know would help you better manage your pastures and extend the grazing season? Could you find the time to better market your livestock or end products of meat or milk? Could you even find the time to take a family vacation?

Time is the most valuable resource in your life. You need to make decisions about how to use it most effectively.

Perhaps after time, land is the most valuable resource you manage. Land provides the base of the solar panel we manage to capture free solar energy and turn it into a salable product. When we generate a salable livestock product from standing forage the animal has harvested themselves, we have done so at a much lower cost than when we harvest the forage with labor, iron, and dead dinosaurs. Mechanical harvest takes low cost forage and turns it into high cost feed.

To be in a profitable business, you must know your cost of production.

Chapter 3
What did animals do before there was hay?

If hay has been part of our culture for only 4000 years, what did animals do for the hundreds of thousands of years before that?

Livestock have been domesticated for at least 10,000 years in southwest Asia. Archeological evidence from Iran clearly shows domestication of goats as early as 8,000 BC. By studying the developmental stages of bones found in the Ganj Dareh settlement in the Iranian highlands, scientists have been able to show the systematic slaughter of young male goats while also simultaneously showing prolonged age in female goats. Definitely a domesticated herd of goats.

The highlands of Iran are characterized as a high desert region. Goats are the type of animal who could survive browsing on twigs and dead stems during the cold, dry winters of this region. There would have been no need for hay in this environment. Another interesting piece of history is this same general region is also where alfalfa was a native plant. Alfalfa is generally considered to be the first plant domesticated solely for use as a forage. Early on, alfalfa was only a grazed crop, but we know that by Roman times alfalfa was being harvested as stored fodder. In more northern climates, as cattle and sheep were domesticated, the need for winter forage would have been much greater. Hence the hay stacks in Romania twenty centuries ago.

We can look at wild animals in colder climates to learn how they deal with winter without human intervention. Unfortunately we see increasing numbers of elk and deer becoming welfare bums preying on hay stacks of farmers and ranchers. In some locations, thousands of elk are being fed hay in the winter by well-meaning humans to stave off starvation. Concurrently, we see declining survivability of these herds as disease increases and they lose their ability to fend for themselves. In North America, the bison in Yellowstone National Park and the bighorn sheep in more remote parts of the Rockies provide better clues for how to make our domestic livestock less dependent upon hay.

The first critical factor to recognize is no ruminant animal in the wild gives birth before green forage has begun to grow in the spring. The nutritional demands of breeding and lactating females are very cyclic and very predictable. Lowest nutrient demand occurs after the dam has weaned her offspring and lactation has ceased. Highest requirements are when the female has reached peak lactation.

In the case of large ruminants, peak lactation and rebreeding occur fairly close together. We generally consider beef cows to reach peak lactation 45 to 75 days after calving. For a cow to remain on a twelve month calving interval, she must rebreed between 80-85 days after calving. This places tremendous nutrient demand on the female, particularly two and three year old heifers and cows who are still growing themselves. The case is very different for small ruminants that reach peak lactation in just a few weeks, but do not need to rebreed until several months later.

Wild animals wean their young naturally as milk production declines when forage becomes more limiting in late summer and fall. Even though antelope, deer, elk, big horn sheep, and bison stay together as family groups through the winter, and frequently much longer, the yearling offspring will not resume suckling their dams when the new babies are born.

Bob Jackson, a long time backcountry ranger at

Yellowstone and a bison behavior expert, says he has never seen a yearling bison nursing on a bison cow in all of his years observing the herds at Yellowstone. We sometimes see this in beef herds because of the high milk producing abilities we have bred into some of our cows and the lack of the natural weaning process.

Wild animals do not produce excessive amounts of milk. Natural selection has brought them to a lactation level that is consistent with the foraging opportunities available in their environment. We humans, on the other hand, have taken our livestock to levels of lactation far beyond what the natural environment can provide, hence the need for supplemental feeds. After all the years of buying bulls with high milk EPDs, most cattleman are coming to realize the high cost of keeping the daughters of those bulls. Their genetic ability for milk production far exceeds the carrying capacity of their forage resources. Hence the need for hay or other expensive inputs.

Large ruminants with gestation periods of nine months or so key in on the equinox for estrous and fertility. Bison and cattle have their peak breeding cycles in September in the Northern Hemisphere and March in the Southern Hemisphere. Short gestation ruminants like sheep or goats key in on the solstice. The natural fertility of northern sheep breeds is greatest in mid-December. Both of these periods of high natural fertility for long or short gestation animals result in birthing seasons in late spring and early summer. All of the bison, deer, elk, antelope, and bighorn sheep in a particular area give birth in a remarkably short span of time. The simple reason is they are much more closely tied to the key signals in nature than are our domestic livestock.

It is interesting to note how both plants and animals are tied to changing day length throughout the seasons. Photoperiod is the key that tells plants when to flower and when to start storing carbohydrates for winter survival. Photoperiod is the key to telling a ewe when to resume estrous. Photoperiod changes are extreme at far northern and southern latitudes while they become minor the closer we come to the equator. Livestock breeds developed in northern Europe have much stronger photoperiodic response than do their tropical counterparts.

The bottom line here is moderate milk producing females giving birth in sync with natural forage production peaks are much more likely to work in year-around grazing systems than will high milk-producing super cows calving in the middle of winter.

Even if we get our livestock to breed and birth in sync with nature, we might look at our pastures and conclude there is nothing worth eating out there. Once again we can look at nature for our answers.

Our pastures are often monocultures or very simple mixtures with a couple of grasses and legumes. Most natural plant communities have a wide diversity of grasses, legumes, forbs, shrubs, and trees. Wild animals may use all of these feed resources over the winter. The relative amounts of protein, energy, and minerals vary across species. To maintain a balanced diet in the dormant season, plant diversity is an important consideration.

It may not be necessary to have all five classes of plants listed above in every pasture, but it is certainly valuable to have more than one. Stockpiled tall fescue is the closest thing to a monoculture I have seen work for nutritious overwintering of livestock. But we found it was even better with other grasses and legumes in the mixture.

Healthy native rangelands in the American West provide much better winter nutrition than do the seeded pastures of crested or intermediate wheatgrass. Choosing which of your grasslands to use for winter grazing is another component of success. Later chapters will explore the many options we have available for winter grazing.

The first step in planning for successfully kicking the hay habit is look to nature for the best example of low-cost winter grazing.

Chapter 4
The power of year-around grazing

For a long time, winter feed costs have been the most expensive part of being in the livestock business. Raising livestock has been an unprofitable business to be in for many of the participants. Numerous studies have shown winter feed costs to be the largest single factor for determining the profitability of a cow-calf operation. The logical conclusion we can draw from this string of information is the quickest way to put profitability back into a cow-calf operation, or most other livestock enterprises, is to tackle winter feed costs head on.

When I first began working at the University of Missouri-Forage Systems Research Center (FSRC) in 1981, grazing stockpiled tall fescue pasture was already a standard management practice, but what we didn't realize was how far we were from the real potential of stockpile grazing. At that time, the stockpiled pastures were either set stocked for as long as it would last or strip grazed with electric fence moved every two to three weeks. We considered the winter feeding period to be about 130 days, roughly from the beginning of December to early April. An acre of stockpiled tall fescue at this time normally provided 45 to 50 CDA (cow days per acre).

To take a dry cow through the winter would require 2.5 to 3 acres. The general opinion around both the farming community and the university was, while stockpiling was a useful practice and could save a little money every winter, it was entirely unpractical to try to take a cow all the way through the winter just on pasture. The rationale was no one could afford to set that much land aside just for stockpiling. So no one really took the idea of year-around grazing very seriously in those days. And I was one of them.

We thought we would always need to make some hay because you could obviously harvest more hay per acre than you could take off as stockpile. What did Will Rogers say? "It ain't what you don't know that hurts you, it's what you know that ain't so that will get you." It took awhile but we learned a well-managed acre of pasture would almost always outproduce the same field managed for hay.

For the first several years I was at FSRC, I accepted the dogma of year-around grazing being impractical, rather than just asking how can we get more from an acre of stockpiled forage. Like so many other things I learned about grazing, I learned it first at our home farm.

Many Midwesterners will remember 1988 as one of the driest years in memory. We had made a new 12-acre pasture seeding on our home farm that spring. We watched it come up beautifully in April. A veritable salad bar of orchardgrass, brome, timothy, red clover, and birdsfoot trefoil. And then we watched it wither to dust by June. But what did come was some foxtail, barnyardgrass, and crabgrass, along with some other weeds and a little bit of annual lespedeza. We didn't get much rain and the weeds never really got very big, but we grazed nearly 100 ewes with lambs on that 12 acres for over 60 days. The driving factor was daily rotation.

This was our first experience with management at this level of intensity. We were doing mostly three to four day rotation stuff, which was pretty intensive for the neighborhood at the time. Here we were in a drought, grazing weeds, but seeing our lambs growing like they never had before. Even without rain, the weeds still grew a little bit and, with some rest, they gave us the grazing we needed to make it through the drought. As an agronomist I mostly thought about daily rotation just giving us more rest days and the ability to grow grass. I really didn't think all that much about what was going on with the nutritional side of the equation.

In every pasture, there is excellent nutrition available. There is also less than ideal nutrition available. A skilled grazing animal seeks out the best bite it can find. After they have taken away all the excellent forage, they must settle for something less.

When that next level is gone, they must eat stuff they really didn't care for on the first rounds and their intake rate will go down as will the nutrient density of each bite. Grazing animals are always sorting through the salad bar and creaming the best bits of food. At that time I didn't really understand the value of putting the good, the bad, and the ugly all into the rumen together. Daily rotation took away the opportunity for selective grazing and thereby moderated the rises and fall in daily intake that we saw in longer grazing periods.

When winter rolled around, we had a lot less hay in the stack, as well as being short on stockpiled pasture. At the home place, we employed daily strip grazing on the stockpile and found we got just as far into the winter before feeding hay as we did in a normal production year. At FSRC we started moving fences every couple days rather than every couple weeks and found we grazed much farther into the winter than in most years. The drought-induced, half-crop hay yield turned out to be all the hay we needed. Necessity is indeed the mother of invention.

Another observation was the stock, cows and ewes, both seemed to be in better body condition in midwinter when the stockpile was finally used up than they would normally be if they had been eating hay at that time. We had always expected cows to be losing body condition towards the end of the stockpile period, but this year they were noticeably more robust. At first I attributed it to the known fact that drought stressed forage can be significantly more nutrient-dense than forage grown in a wetter year. That was a plausible enough explanation so I didn't look any further.

After the 1988 experience, we began doing a lot more work at FSRC evaluating more intensive systems and trying to understand why they were so beneficial for animal well being. On the home farm, we shifted to daily moves for both the cattle and sheep during the growing season as well as winter. Once you see the benefits of such management in your operation, it is hard to go back to anything less intensive. Yet, I still believe more intensive management actually has greater economic potential in winter grazing than it does during the growing season. Why?

In the active growing season, and particularly in the spring, the alternative to a bite of cheap grass is another bite of cheap grass right next to it. Why should you even bother to manage more intensively to produce more forage when you already have an excess? On the other hand, during winter the alternative to a bite of low-cost stockpiled pasture is high-priced hay. There is little or no pasture growing in the winter dormant season, so when whatever you have stockpiled runs out, the expensive days of hay feeding begin.

Any winter feed in the northen half of the USA needs to be produced during the growing season. One of the strongest arguments I can make for using MiG (Management-intensive Grazing) in the summertime is to be able to grow feed for the wintertime. Uncontrolled grazing throughout the growing season will leave you with no opportunity to grow stockpiled feed. The more intensively you manage during the summer, the more winter feed you are likely to stockpile. In the winter we can use strip grazing to manage the feed supply. Intensive strip grazing is a powerful tool for getting the most out of your winter pastures.

Where is winter grazing practical:

Stockpiled perennial pastures can be used throughout the winter in much of the central and southern USA. In the northern third of the USA and into Canada, snow depth may limit usefulness of these pastures for much of the winter. Where snow becomes a limitation, pastures not utilized before deep snow cover can often be partially salvaged with spring grazing.

I often use primary highways to delineate different management zones within the USA. Winters are usually open enough to allow grazing most of the winter in the area south of a line along Interstate 80 from Reno, NV, eastward to Chicago, I-65 from Chicago to Indianapolis, and then south of I-70 eastward to the Atlantic Coast. There will be pockets north of this line with open winters as well as some areas south of the line where snow depth may become limiting on a regular basis. But for most areas south of that line, snow should not be an insurmountable obstacle except at higher elevations.

South of I-70 all the way across the USA, winter annual pastures can play an important role in your winter grazing program. Winter annuals like annual ryegrass, the cereal grains, and the brassicas may grow most of the winter south of I-70. Annuals may be used in some areas north of the line, but their growth will be primarily only in autumn and then again in the spring.

In the deeper snow areas of the Plains and Prairies, swath grazing is a useful tool for making feed more accessible to livestock. Using taller growing summer annuals like corn and sorghum can put forage above the snow level and allow grazing to continue in many parts of the northern USA.

Where is year-around grazing likely to be the most challenging? In my view it is the Great Lakes region and New England. This is consistently deep snow country and the forage options may be more limiting. This is an area where dairy farming still predominates. The higher nutritional needs of lactating animals put more stress on the system.

Another region where year-around grazing is difficult is the coastal regions from northern California to British Columbia. The problem here isn't snow depth. It is the continuous rainy weather that keeps the soil so saturated the ground cannot physically support the weight of the animals.

In most of the USA, year-around grazing should be the norm, not the exception.

Strip grazing for high utilization:

When we first went to more frequent fence moves on stockpiled pastures, our primary motivation was increasing harvest efficiency. All you have to do is think about what else a cow does on her dinner plate and it is easy to understand why more frequent moves lead to higher grazing efficiency. When an animal has the opportunity to eat, sleep, wander, and defecate all on the same space, some of the feed resource is inevitably going to be wasted.

One of the keys to reducing forage wasted by fouling and trampling is to simply limit the amount of time the pasture is exposed to the animals.

How much difference does the length of the grazing period really make? We conducted a four-year study at FSRC in the early 1990s that looked at the effect of frequency of stockpile allocation on pasture utilization and animal performance. The treatments provided either 3, 7, or 14 days worth of standing feed in each allocation. The cows were weighed and condition scored every 28 days so we knew what their daily forage requirements should be. The pasture was measured for yield. Quality samples were taken. Each allocation was based on the needs of the cows and the availability of the pasture.

What we found at the end of the study was very enlightening. Cows given a 14-day feed strip, were moved on average on the 9-10th day. The pasture looked like it was gone and cows were complaining.

Closer examination found forage left under or around manure piles, matted down grass frozen to the ground where they had bedded, and some of the less desirable forage left ungrazed. Cows on 7-day strips were usually moved after five days based on the same criteria. Cows on 3-day strips, on average, stayed the full three days. At the end of the season we had taken 40% more grazing days off the 3-day strip compared to the 14-day allocation. Animal performance was similar across all treatments.

A 40% increase in low-cost winter feed supply is pretty significant. That means 40% fewer acres need to be set aside for stockpiling leaving us with greater flexibility in our summer and autumn use. At the time of the study (1991-1994), the cost saving from grazing stockpile rather than feeding hay was about 75 cents/cow-day. In 2008, that savings was well over a dollar per cow-day across most of the USA. This is one of the quickest ways to cut out $100 per cow in annual expense. This is one of the quickest ways to put profitability into a cow-calf operation.

Strip grazing for nutrition:

As the years went by and we continued to see improved animal performance and body condition from more intensively managed winter pastures, I came to realize the nutritional impact of strip grazing was every bit as significant as the enhanced utilization. Working with Dr. Ron Morrow and Dr. Fred Martz and their graduate students, we began to look at the daily intake level and the nutrient content of the grazed forage managed with different grazing periods. Intake and nutrient density went down with each subsequent day.

The difference in intake between the first and seventh day of a week-long grazing period was nothing less than shocking. We were putting animals on a roller coaster diet with grazing periods in the 5-10 day duration. That yielded inconsistent performance. When put on a daily rotation, intake and nutrient content was very consistent from one day to the next yielding consistent and predictable performance.

Grazing periods longer than ten days during the growing season actually give better performance than weekly rotation because the animals start selecting new regrowth as a supplement to their diet. Not good from the plant's perspective as removing the new regrowth could weaken the plant, but a good nutrient boost for the livestock. In the winter, there is no new regrowth to act as a supplement so the selection of good quality material in the first several days of the grazing period left them with little to choose from in the following weeks.

It doesn't take a lot of green material or cured leaf to provide enough protein supplement for the rumen microbes to digest more mature stockpiled forage. Daily strip grazing put a consistent diet into the rumen that animals could do relatively well on, hence the improved body condition we were seeing. Protein supplementation of ruminants, however, does not need to occur on a daily basis. Supplementing every three days does about as well. Strip grazing on a three or four-day basis may not give quite the performance that daily allocation does, but it puts the animal in a very acceptable position to be able to perform well on stockpiled forages.

Cost savings of winter grazing:

Every farm and ranch has a different set of costs associated with that particular operation so it is very difficult to say, "You will save $97.38 per cow every winter if you graze rather than feed hay." However, I will say unequivocally you will save money if you follow the basic guidelines outlined in the remainder of this book.

Not all sections may apply to your particular situation and you may have unique challenges not addressed here. Progress is made by gleaning information from many sources and adapting a variety of ideas and techniques to your situation.

Chapter 5
Know your forage opportunities and challenges

One of the first things to understand about year-around grazing is it requires a lot of forward planning and effective management to accomplish the goal. Having a sound understanding of the forage production potential and opportunities on your particular farm or ranch is one of the first steps in building your plan. You will also have a set of challenges to having a year-around grazable forage supply. There is no simple one-size-fits-all winter grazing program. It's all about managing the resources you have available.

There are five basic forage resources discussed in this book: 1) Crop residues, 2) Cool-season perennial pastures, 3) Warm-season perennial pastures, 4) Native rangeland, and 5) Annual pastures. All of these will be covered in greater detail in later chapters. It is conceivable that a single farm or ranch may have all of these resources available, while at the same time, another farm or ranch may only have one resource with which to work.

First think about the general area where you live. What types of farming or ranching take place there? What kind of pastures are commonly used there?

While we often chide mainstream agriculture for being inefficient, tunnel-visioned, or just downright dumb, there are usually sound reasons why certain farming practices and crops developed in particular regions. The basic reason is because at one time or another it made economic sense to implement that practice or grow that crop. In the 1960s it made perfect economic sense to put all the dairy cows in barns and haul all the feed to them mechanically. It was based on the relative price of milk compared to costs for equipment, fuel, and labor at that time. When those relationships change, management needs to adapt to the changing conditions.

Unfortunately, in farming and ranching, it usually takes at least the generation once removed from the time of economic change for the adaptation to take place. That is why most farmers and ranchers always seem to be just hanging on by a thread. They are usually managing based on the economic climate of their parent's peak years, not the times in which they are living.

When it comes to pasture, look at what has been grown in your location for years. If it has been growing there for years it means it will probably continue to grow there for years. These well-adapted pastures or crops should form the foundation of your grazing program.

Does that mean we need to limit ourselves to these few choices? Absolutely not! What it means is you don't need to tear up everything you have and start from scratch with some theoretical year-around grazing scheme drawn up by someone who lives on the other side of the world or even just the other side of your state or province. Working with what is already proven to be well adapted is just a good starting point.

Planning for year-around grazing begins with knowing when you typically have ample pasture available and then identifying the times of forage deficit. Generally speaking, the longer your growing season, the shorter the time period you need to fill in with alternative forage sources. It is also important to be aware of how forage quality changes through the seasons. The next chapter will delve deeper into understanding forage quality.

I like to plot a line or draw a chart of when different forages grow in a particular location. This graphic helps you better see the peaks and valleys of your forage supply. Just writing down two dates as beginning and ending points of the growing season for a particular pasture type does not have the visual impact of a chart or drawing.

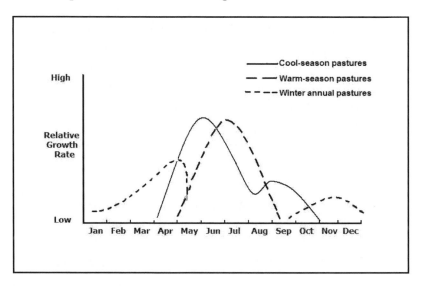

Here are a couple of examples:

Looking at the figure above we can see the usual pattern of abundance of forage in spring and summer and deficiency over winter. There is a lot of overlap with the cool-season and warm-season pastures through the peak of the growing season. The winter annual pasture provides some fall pasture as well as earlier in the spring, but the yield potential of the annual doesn't appear to be as great as the perennial pastures.

The traditional strategy for dealing with these periods of surplus and deficiency has been to cut hay in the times of excess and feed it out in times of deficit. This is what we're trying to get away from. The hay alternative has simply become unaffordable for most livestock operations.

This little bit of information is enough to tell us we either need to figure out how to put more of that spring and early summer growth potential into winter feed or we need to adjust our stocking rate to create a lot more animal demand in the summer and much less in the winter. Stockpiling forage during the active growing season and deferring use until winter is one way of moving forage from one season to another. But shifting forage availability through stockpiling always comes with the price of lower quality forage in the dormant season. Animal demand can be increased or decreased by timing of the calving season or adding additional stock with a custom grazing enterprise. We'll explore these and other options in subsequent chapters. The point is there is always an alternative.

The bar chart is another way of showing when pasture is usually available.

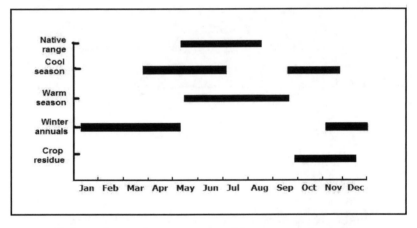

It is easier to draw a reasonably accurate bar chart on the back of an envelope or cardboard box on your tailgate than it is to draw the growth curves.

I like the bar chart as a starting point. Just draw a line for the time period a particular pasture is growing. Anywhere you have three bars overlapping, you probably have reliable pasture almost every year. Two bars overlapping means you're generally safe, but you should develop an alternative strategy for that time period in case of drought, flood, locust plague, or whatever else might come down the road. Where there is only one bar or no bar covering a particular month, that is where your management efforts really need to be focused because that is where hay becomes the likely alternative.

These two illustrations are very simple charts as they deal only in broad generalities. Within each of the categories shown there are likely to be pastures with differing species composition that provide grazing either earlier or later than the generality might show. It is important to get to know each of your pastures individually. I think this individual familiarity is important whether we are talking about a farm consisting of 40, 400, 4,000, or 40,000 acres.

Keeping records on an individual pasture basis provides you with a great arsenal of management information. Everything from knowing how many cows it will carry in August to the relative cost per grazing day can be determined if you have consistent records. You will find some pastures have decided cost advantages for using as winter pasture compared to just summer grazing. Maybe you have a gut feeling about some of these things, but sound baseline data will usually get you closer to your goals more quickly than just working on a hunch.

Neither of these graphs tell us anything about the relative acres of each of these pasture types. If 90% of the pasture is in the cool-season perennial category, just having three bars running across a month doesn't really mean you're covered.

Number of grazing days from each pasture type is the key piece of information. Total grazing days is the product of grazing days per acre for each pasture type times the number of acres for that pasture. We can then break it down by month or even bi-weekly intervals to have a better understanding of exactly when we are likely to have forage available from a particular pasture type.

The tool we use to define our pasture supply in detail is a pasture inventory. As the name implies, a pasture inventory is an ongoing record of what is available for grazing on the farm or ranch at any point in time. The inventory can be used for forward planning based on past records from your own pastures.

While the pasture inventory is a year-around tool, it

may be the most helpful when it comes to planning your winter grazing program.

The unit of pasture yield I most commonly use is cow-days/acre (CDA). This is conceptually similar to using standard animal unit days (AUD), dry stock equivalent (DSE), or any of the other terms used by various authors and agencies. It just happens that when I walk into a pasture the first number that comes into my mind is CDA.

What is a CDA? It is the amount of forage consumed by one cow in a 24-hr period (a day). Your CDA is likely to be different from my CDA due to differences in cow size. We can always convert to standard animal units to compare our records.

On my farm, though, the only thing I'm really interested in is how many days can I feed my cows on my pasture. The unit of measure you choose to use need only be representative of your stock on your operation. This process is designed to help you more effectively manage your resources, not give you some sort of bragging rights at the coffee shop.

If you are a sheep producer, you might want to use a ewe-days/acre (EDA) as your standard measurement. If all you do is run stockers, use an SDA (stocker-days/acre).

Yes, if you have a diverse, mixed species operation, you may want to convert everything to standard animal units or determine what the appropriate conversion ratios are for your particular stock. Obviously if your stocker program typically takes cattle from 400 lbs to 700 lbs your SDA will be different from a neighbor who grows cattle from 600 to 900 lbs.

So why don't I just insist we all use the standard animal measurement to avoid the confusion? Because confusion only arises when we're trying to compare from one operation to another. Within the bounds of your farm or ranch, use the unit that makes the most sense to you.

If we converted everything to standard animal units when we were at a conference or grazing network meeting, when we got back home we would all have to convert it back to something we could relate to anyway. With 95% of your use of the inventory relevant only to your home operation, use what you're comfortable using and save the standard animal unit conversions to those rare occasions when you have to talk to someone else.

The better you understand your forage resources, the better job you can do of managing those resources.

Chapter 6
Conducting a pasture inventory

Using a pasture inventory to monitor your forage supply is one of the best tools to start planning your winter grazing program. The inventory allows you to assess how much standing forage you have on hand for winter grazing. Counting bales in the stack yard is simple to do and most ranchers regularly count bales. Counting grazing days in a pasture is a little more challenging but it can also be done effectively with a little training and experience.

There are different ways to conduct an inventory. To a certain extent, it depends on how hung up you are on processing numbers.

I'm going to outline three different approaches to estimating pasture productivity. The order of presentation is from most complex to simplest.

Reading that, a lot of readers might choose to just jump ahead to the third and simplest process. However, I think it is important to understand how these processes evolved, so I encourage you to read the entire chapter and then decide which process you prefer.

The pasture dry matter yield method:
In the research business, we generally measure pasture yield by clipping a multitude of standard sized frames or quadrats (usually .25 meter squared), weighing and drying the

clippings, and then calculating pasture yield per acre from these small samples. This is a time consuming and expensive practice but it yields fairly accurate results. Over the years we began to always take a height measurement in every quadrat we clipped. The result of this combined sampling was the conclusion that the taller a pasture is, the more of it there is. This is not rocket science!

There are a couple of other details that need to be considered other than just pasture height.

Denser sods yield more forage per square foot than sparse sods. So not only do we need to measure height, we need to also estimate stand density. The correlation of height to yield is fairly good as long as pastures have at least 60% ground cover. If much more than half the ground is bare, the shape of how plants grow changes and the height:yield relationship breaks down.

The other factor is pasture type.

Some forage species just yield a lot more per inch of height than do others. Kentucky bluegrass is a very good example. Because of its fine leaves and dense sod formation, six inches of bluegrass may contain as much forage dry matter yield as twelve inches of orchardgrass.

The following table was developed while I was at the University of Missouri and it combines the three factors of height, stand density, and pasture type to predict forage yield. While it is far from perfect, it is a good starting point for doing a pasture inventory.

When evaluating stand density of bunch type native grasses, look at canopy cover when the grass is about ten inches tall rather than cover at the 3-4" level.

In this inventory process, we record a yield estimate for each pasture in units of lb/acre of available forage. We would then go through a process using animal weight, expected intake rate, and target utilization rate to calculate expected number of grazing days per acre.

Table 1 Estimated dry matter yield in pounds per acre-inch for several pasture types and stand conditions.

Pasture	Stand Condition		
Species	Fair	Good	Excellent
	(lb/acre/inch)		
Tall Fescue + N	250-350	350-450	450-550
Tall Fescue + Legumes	200-300	300-400	400-500
Smooth Bromegrass + Legumes	150-250	250-350	350-450
Orchardgrass + Legumes	100-200	200-300	300-400
Bluegrass + White Clover	150-250	300-400	450-550
Mixed Pasture	150-250	250-350	350-450

Here is the process:

Step 1. Identify the pasture and subdivide the sampling area if needed. If there are notably different areas of productivity in the pasture, sample them as individual areas.

Step 2. Measure or estimate the area of each inventoried pasture or paddock. We will be estimating available forage per acre, so area of each pasture is essential to know.

Step 3. Make a series of height measurements across the pasture. Take a minimum of ten measurements. Taking up to 20 measurements may improve the estimate, but sampling beyond 20 has little value and takes a great deal more time. The pasture height should be read near the top of the canopy of free standing plant leaves. Do not measure seed stems and do not pull lax leaves to their maximum height. The height we are looking for is the canopy height below which 90% of the forage yield occurs.

Step 4. While taking the measurement also estimate stand density. An excellent pasture has at least 90% of the soil surface covered by desirable forage plants when it is at three to four inch height. A good pasture has 75 to 89% ground cover and a fair pasture has 60 to 74% ground cover. Pastures with less than 60% ground cover cannot be accurately estimated using this process.

Step 5. Choose the appropriate pasture type from Table 1 based on pasture species and density. Identify the appropriate yield per acre-inch on the table and multiply that number times the pasture height you measured.

Example: If you determined the pasture was best described as an orchardgrass-legume mixture and the stand density was about 80%, the yield multiplier would be 200 to 300 lb/acre-inch. This is quite a range and this is where your good judgement has to come into play. If the pasture appears vigorous and healthy, use the higher factor. If the pasture looks stressed or spindly, use the lower factor. If we measured a vigorous orchardgrass-legume mixture at ten inches, the estimated dry matter yield would be about 3000 lb/acre.

From here we would use target utilization rate and intake target to calculate the expected daily carrying capacity using the standard stock density formula:

Stock density = $\dfrac{\text{(Available forage X Temporal utilization rate)}}{\text{(Daily forage intake X # of days in grazing period)}}$

Assuming forage availability of 3000 lb/acre, target utilization of 50%, and 2.5% of bodyweight intake level, we can make the following calculation:

$$\frac{3000 \text{ lb forage/acre X .5 utilization rate}}{.025 \text{ lb forage/lb liveweight/day X 1 day}} = 60{,}000 \text{ lb livewt/ac/day}$$

Remember when we're doing an inventory calculation, the grazing period is always one day. Given a target stock density of 60,000 lb/acre, we can convert this to cow-days/acre (CDA). If our cows weigh 1200 lb, the inventory value would be 50 CDA.

Cow-days per inch method:

The inventory process can bypass the above method and

go directly to grazing days or cow-days per acre. Once we have learned that pastures of certain heights and densities produce a certain amount of forage, the next conclusion we can reach is pastures of certain heights and densities can feed a certain number of animals for a day.

What we have found is an inch of grazed forage can be expressed as cow-days per acre-inch for different density of pasture. This system is usually tailored to your particular livestock operation. The values listed below are based on 1200 lb cows with calves at their side. Dry cows have slightly higher carrying capacity per inch. An inch of grazed grass will not feed as many 1400 lb cows. This system can work very well once you learn the criteria for your operation. The most important step of this method is to make the estimate and then record what actually happens. That is how you can build your own calibrations.

Here are the basic estimation criteria:
One inch of thick grass provides about 15 cow-days per acre.
One inch of average grass provides about 10 cow-days per acre.
One inch of thin grass provides about 5 cow-days per acre.

If you can't quite decide whether a pasture is average or thick, feel free to use a value in between the two, such as 12 cow-days per inch. Learn to make the adjustments based on specific conditions and this system can work very well.

Here is the step-by-step process:
Step 1. We look at the grass and say, "This is average grass" which gives us a cow-day/inch factor of 10.
Step 2. We measure the height to be 10".
Step 3. We would like to leave a 4" residual, so 10" - 4" = 6" to be grazed.
Step 4. 6 inch grazed X 10 cow-days/acre-inch = 60 cow-days/acre.
Step 5. The inventory value we enter is 60 CDA.

Some readers might be concerned at this point because in the first example our hypothetical pasture calculated out to 50 CDA while the latter method gave a value of 60 CDA. These are only estimates! Neither of them is likely to be the "TRUE" answer.

As you learn to use the CDA/inch method, you'll learn to make adjustments for the little bit thinner or thicker stands, for drought stress vs. lush pasture, and so on. Remember we were using estimates of 5-10-15 CDA/inch for the thin-average-thick pastures, respectively? Why couldn't it have been 4-8-12 or 6-12-18? Because most people can multiply by 5, 10, or 15 in their head!

The range for "average" pasture was 75 to 89% ground cover. Could you tell whether the pasture was really 72% or 78%? No! But 80% (our original estimate) is closer to 75 than it is to 89, so you could adjust the expected CDA/inch from 10 to 8. Now with 6" to be harvested and a CDA/inch factor of 8, the predicted yield is now 48. Wow! That is only 4% variance from the calculated value using actual measured height!

These are only estimates! Don't get hung up on the actual number.

Direct cow-days estimate:

This approach requires training your eye by estimating how many grazing days are in an area, grazing the area, and then comparing how many days were actually achieved to your pre-graze estimate. With practice you can become very proficient at inventorying pastures very quickly. The key is to always compare the actual grazing to your estimate. This method is particularly useful on rangeland where changes in species composition and stand density are common.

In this method, we go to every pasture and just visually estimate how many CDA are standing in the field. When I walk into a pasture, the CDA value is the unit of measure that first comes to my mind. Even though I spent many years in the

research business and measured thousands of pastures for lb/
acre yield, I do not automatically look at a pasture and think,
"Oh, this is 3278 lb/acre."

I don't think, "Oh, there are ten inches of pasture here, I
want to leave four inches, so that is six inches to be removed
and it is an orchardgrass-legume pasture that is kind of thick so
the CDA/inch factor is about 12 so the yield of this pasture is
72 CDA."

I just look at the pasture and 70 CDA is what I see. The
variance of my estimate from the actual grazing record is
usually under 10%. This is the product of doing this for a lot of
years on a lot of different types of pasture. This is where I
would like you to be in a few years.

The key is to make an estimate every time stock go into
a new paddock and then record what actually happens. This
method is simple and it works.

How to use the data:

If you conduct a pasture inventory on a regular basis,
you always have a good estimate how much grazing is available
on the farm or ranch. This is particularly important to know if
you are facing drought or planning your winter grazing program. I recommend conducting an inventory bi-weekly during
the growing season for high rainfall or irrigated pastures. A
monthly inventory may be adequate on rangeland due to the
slower growth rate.

An inventory during the growing season gives you a
way to track whether your forage supply is increasing or decreasing. This becomes a particularly useful tool after you have
collected several years of data. For example, if your July 1
inventory shows you have only 80% of the available forage or
grazing days you have averaged over the previous several years,
you know you are headed for trouble and can start adjusting
management or animal numbers to compensate for the shortfall.

Most producers don't recognize they are in trouble until
it is too late to correct without excessive input expenditures or

financial loss due to forced sales. Maintaining an ongoing pasture inventory is one of the best ways to know when you need to make adjustments in stock numbers and type.

Going into the dormant season, an inventory near the end of the growing season gives you a very good estimate of how many grazing days you have available during the dormant season. This can let you plan stock number adjustments based on available forage. Many producers find themselves buying tons of hay in the middle of winter or a dry season because they really didn't know how much pasture they had available.

As the dormant season progresses, conduct an inventory at least monthly. This serves two main purposes. Obviously it lets you know how many more days of grazing are available, but more importantly it monitors the loss of standing forage due to weather conditions or unplanned wildlife use of your pastures. The inventory done at the end of the growing season tells you what you could have available under the best of conditions. The monthly inventories keep your feet on the ground and keep you informed of necessary management adjustments if the forage is disappearing faster than expected.

Knowing what the natural loss in standing forage is over the winter months lets you compare different pastures for their suitability for use at different points throughout the winter. Some pastures should be used as early as possible while you will find other pasture types or locations weather much better and can be deferred for use in late winter.

If you find the rate of decline in standing crop to be more severe this winter compared to past years, you can start making adjustments to compensate for the more rapid deterioration. This might be selling some stock, ratcheting down your strip grazing program, feeding a supplement to stretch the available forage, or any number of other possible adjustments.

I use a spreadsheet program to keep track of the pasture inventories and available grazing days. Each pasture and pasture subdivision that was inventoried is entered on a separate line with corresponding data for number of acres, expected herd use, and planned dates of use. When the pasture is grazed, the actual use numbers are entered and compared to the projections. If I find that every pasture is providing fewer grazing days than projected, I know I need to adjust my estimates downward. Maybe the cows are bigger than I think they are or maybe the pasture wasn't as dense as I thought. Most producers like to think they have 1100-lb cows and thick pastures. In reality, most have 1300-lb cows and thin to average pastures.

Comparing the actual data to the projections is a reality check that should help you calibrate your estimates. This system works just as well for 1300-lb cows and thin pastures as it does for 1100-lb cows and thick pastures.

You just have to recognize that is what you have.

Chapter 7
Choosing the optimal breeding and birthing seasons

Because your livestock have their highest nutrient requirements between birthing and breeding, it makes sense to plan these important dates to coincide with times when you have ample pasture available. There is not a universal date that is right for every location and forage environment.

There are many experts who recommend calving or lambing when the wildlife in your area are having their young. In natural rangeland environments, this is a sound recommendation. The deer, elk, and other wild ruminants are very in tune with the forage resources and weather conditions where they live. They have to be in order to survive. It is an extremely rare event for fawns or elk calves to be killed by inclement weather. These creatures only get one shot at propagating their species so they better get it right.

For many parts of the USA, those optimal birthing dates fall in May and June. In the South, they may come a little earlier while farther North, they may extend into July. When we lived in Missouri, we would see the first fawns at the very end of April, but most came in mid-May.

Here in Idaho, the antelope are the first to give birth and we see them in early to mid-May. The elk and deer come from mid-May to early June. I remember one trip to Yellowstone a number of years ago when we saw a buffalo calf being born on the shoulder of the road in the last week of May. There were already a number of calves on the ground by that time. Another interesting trait of wildlife populations is in any given area, the majority of babies all come within the span of a single heat cycle for that species. As with any herd, there are some early ones and there are the usual tailenders, but 80%+ will be born in a single heat cycle. That means they were all bred basically within one heat cycle.

What are the keys that let wildlife know when they should be breeding? Photoperiod or day length is the key factor while changing ambient temperature is another, but less important, factor. Depending on the animal species, the key may be a little different. With bison having a 9.5 month gestation and big horn sheep at six months, but giving birth at approximately the same time period, they must be influenced by different factors of day length.

Large ruminants with longer gestation periods are tied to autumnal equinox when day length and night are of equal duration. The first day of autumn is around Sept 21, after which day length becomes shorter than the dark period.

This is the key that tells wild cattle or bison to begin the hormonal sequence that will lead them to initiate estrous. Open bison cows do not cycle in midsummer. If they did, their progeny would be born in February and would likely not survive the winter storm season. As the biological system works, their calves will almost always be born when the weather is good and fresh forage is on the range.

Smaller ruminants like deer, goats, or sheep have correspondingly shorter gestation periods. Deer are 200-205 days, goats about 150-155 days, and sheep in the 140-145 day range. Each of these species are geared towards a different ratio of daylight to darkness.

Domesticated sheep usually breed around the winter solstice (Dec 21), the shortest day of the year. One of the quickest ways to increase the lambing percentage in your domestic sheep flock is to change from summer breeding for winter lambs to winter breeding for true spring lambs.

This switch is usually good for a 30 to 50% increase in lambing percentage.

Breeds of livestock that evolved in more northern latitudes are the most closely tied to photoperiod while breeds evolving closer to the equator are less rigid in their breeding seasons. The reasons are obvious when we think in terms of the degree of change occurring in photoperiod as we move from the equator towards either pole as well as the likelihood of cold weather being greatest in the higher latitudes and lowest near the equator.

All the natural processes in the endocrine system of wild animals tell them to give birth in the spring when there is typically ample forage quantity and quality available to assure good milk flow and rapid growth of calves, fawns, lambs, or kids. This is a very sustainable production system that we should pay attention to if we want to manage our livestock in a sustainable manner.

The part of the system we haven't talked about yet is the low nutrient demand most of these wild animals have over the winter period when forages are dormant. The coinciding of lowest demand with lowest forage quality during the mid gestation period is no coincidence. This is also a key to survival for native wildlife.

Daily intake required for maintenance as a percent of liveweight is much lower for the wild animals than it is for our domesticated livestock. A wild elk in the Rocky Mountains can survive with an intake level a little less than 1% of her body weight as a maintenance diet. For most of our cattle we expect maintenance intake to be 1.4 to 1.9% of her body weight.

Even the maintenance requirement of high producing cows is significantly higher than the maintenance requirement of a low milk producing cow. No one has ever selected a wild elk for high milk production. If anything, nature has selected for the opposite.

Humankind has been selecting for higher milk production in cattle for a few thousand years now. In doing so, we have gained the ability to produce much more milk or beef in a shorter period of time, but it has come at a higher daily maintenance cost.

While this discussion makes a pretty good argument for "calving in sync with nature," there are times we might want to look at some different alternatives. Those times are when we are no longer dealing with a natural plant community or when other factors override the basic argument for spring calving.

I think the calving in sync with nature argument is very sound for most of the Western, Northern, and Northeastern parts of the USA. The areas where we may want to do something different are in the Deep South and in the tall fescue belt through the Upper South and Midwest regions. Here are some reasons why calving in late spring might not be the best choice for your particular location.

If you live in the tall fescue zone of the USA, late spring calving might be the worst possible choice. Fescue greens up and starts growing earlier in the spring than most other cool-season grasses and, certainly, long before warm-season pastures in that region.

Peak forage production on fescue may be in late April through mid-May so that is when the calving-in-sync with forage supply would tell us to calve. But spring and early summer is when endophyte-infected fescue is at its most toxic level and it's when your cows need to get bred if you have a spring calving herd. April 15 calving requires bulls be turned in around July 7. This is a recipe for disaster with most cow herds in fescue country.

Let's take a quick look at what is going on with the cows and the fescue in a typical spring calving scenario.

We're going to be kind of hard on these cows and make them eat infected tall fescue hay all winter and then go to infected pastures in the spring. This is the worst case scenario for a cow-calf producer in the fescue region. Pregnant cows eating toxic hay over winter will be losing body condition (BCS) right up to calving time.

The effect of declining BCS on the ability of a cow to come back to estrous and rebreed for a 365 day calving interval is well documented. They won't do it without extensive supplementation. As the calves are dropped and pasture begins to grow, the poor girls are sent out on infected pastures.

Bulls go in roughly 83 days after the first calf comes and the temperatures are starting to get a little warm, if not downright hot. Tall fescue pastures have probably gotten out of control with seed heads exploding onto the scene by mid-spring. This is the most toxic stage of tall fescue and you're asking your cows to regain weight, produce milk, come back into estrous, and get bred. You're asking bulls to maintain semen quality in the face of rising heat stress and increasing nutritional stress. This is what we call a lose-lose situation.

Most cattlemen in the Southern and Central USA understand endophyte-infected tall fescue a lot better than they did 20 years ago. We see a lot more legumes and crabgrass interseeded into pastures to dilute the fescue toxicity, we've seen some pastures reseeded to other grass species, we've learned strategic supplementation to keep cows in better condition, and maybe we've even selected for more fescue tolerant bulls. We've made a lot of progress, but the endophyte can still rear its ugly head in a spring calving situation and knock the heck out of conception rates, calf gains, and profitability.

The bottom line with fescue and spring-calving cows is you've got the deck stacked against you.

Why does fall-calving and grazing stockpiled pastures make so much sense with tall fescue?

Location of the endophytic fungus and concentration of toxins in specific plant parts is one big reason. The greatest concentration of toxin is in the seed head and stems while the lowest concentration is in the leaf blade. Tall fescue basically only heads out in the spring so once that flush of stems and seed heads is gone, toxicity is greatly reduced. In a well managed stockpiled pasture, the fall forage is mostly leaf blade and is much lower in ergovaline concentration than spring forage.

Recent research by Dr. Rob Kallenbach at the University of Missouri has shown the ergovaline concentration in stockpiled forage rapidly declines over winter with cold temperatures. Even pasture that starts the winter with fairly high toxicity levels becomes much safer to use by mid-January. In the summer time, the endophyte just keeps getting compounded by rising heat and reduced forage availability.

Contrast this to feeding infected tall fescue hay. Hay is almost always highly toxic because it usually doesn't get harvested until the seed heads are fully emerged and toxicity is at its highest point. The toxin levels stay high in hay and can even carry over into subsequent winters if the hay isn't used in the first winter. Because tall fescue stockpiles so well and retains its nutritive value through most of the winter, it just makes sense to graze it rather than hay it.

Aggravated heat stress is one of the primary symptoms of fescue toxicity. With spring calving cows, breeding almost always occurs as temperatures and stress on the cattle are increasing. With fall calving cows, breeding takes place in late autumn or early winter thus minimizing the possibility of heat stress. It's a lot easier to maintain tight breeding and calving seasons in the fall than in the spring and summer.

Because tall fescue is such an excellent grass for stockpiling as standing winter pasture, we can feed fall calving cows and calves very economically. If the stockpile is properly grown, it can meet the needs of a lactating cow throughout the winter.

While I was at FSRC, we took fall calving cows through the entire winter on stockpiled tall fescue-legume pastures with no supplementation other than salt and minerals. The cows bred at 92% in 45-day seasons and lost less than one BCS unit over winter. Our feed cost for carrying a pair through the winter was under $50.

The bottom line with infected tall fescue is planning your calving season to avoid toxicity and capitalize on its stockpiling quality makes autumn the optimal calving season if fescue is your primary forage base.

Another time to consider calving out-of-sync with nature is when your most productive and highest quality forage is from winter annual pastures. This is the situation in much of the Deep South.

Summer pastures are often bahia grass or lower quality common bermuda. A spring calving situation often has cows on mediocre forage at breeding time and has the opportunity for a lot of heat stress. The same pastures overseeded with annual ryegrass may have the highest quality forage of the season in late winter or very early spring. Because winter weather in the Deep South is fairly mild, calving in January or February can be quite pleasant. Cows get bred in early to mid spring and utilize the summer pastures in late lactation or when they are dry, pregnant cows.

Winter calving in the Deep South is not without risk. There are the occasional storms that drop the temperature below freezing even in central Florida. Not only do the citrus groves suffer, but so can new born calves and thin-hided Southern cows. You should be aware of the frequency of such storms and know the risk level involved.

Another consideration for the Deep South is the possibility of heat-induced embryonic loss. Cows that get bred in April-May and are then exposed to the heat of summer may reabsorb the fetus during the first several weeks of pregnancy. This is much more of a problem with English breeds of cattle than it is for the more common Brahma or Zebu (Bos indicus) breeds. As little as 1/4 Bos indicus breeding can make a difference in preventing early embryonic loss.

Weather conditions at time of calving or lambing is obviously an important consideration. No one enjoys pulling calves in the middle of a blizzard or out on an ice field. It just makes good sense to avoid having calves, lambs, or kids dropping when weather conditions routinely threaten their very survival. These babies are born with summer hair coats and fleece, not winter coats. That in itself is enough to tell us from an evolutionary standpoint that these creatures weren't meant to be born in winter conditions.

Getting in sync with nature is an important consideration except when your pastures are no longer natural.

Chapter 8
Achieving a variable stocking rate

We sometimes speak of the "head slap" moment. It's that moment when you realize you've been thinking about something in the wrong way for a very long time or when something occurs to you that you had never before thought of. Most farms and ranches are stocked based on their summer carrying capacity. That basic stock policy dictates they will have excess spring forage and very little winter pasture. The historic way of dealing with the spring peak and winter valley has always been cutting hay in the peak and feeding it in the valley.

My head-slap moment on our Missouri farm came when I realized the more sensible stocking policy was to stock the farm to its winter grazing capacity and then consume the spring forage peak with additional animals. This is what we mean by a variable stocking rate. It took us several years to figure out just how much additional grazing pressure we needed during the peak season and when the animals should arrive and leave. We used about 60% of our total grazing capacity year-around for the cow herd and then crammed 40% of our annual stocking capacity into the 100-120 days from mid-April to early August.

Pasture does not grow at the same rate every day of the year. No matter where you live, there is variation. The variation may range from a growing season of less than 100 days in some northern latitudes or at high elevation or in some bone-dry desert region all the way up to 365 days in more equatorial latitudes. Even where temperatures may allow for a year-around growing season, variations in precipitation and subtle changes in temperature affect the rate of pasture growth.

Usually the shorter the growing season, the more extreme the peaks and valleys of forage supply. Regions with very marked wet and dry seasons have greater variations in seasonal forage supply. We can seed a variety of pastures to try to even out forage distribution. We can irrigate. We can fertilize. We can stockpile to carry forage from the growing season into the dormant season. We can do all these things, but we will still have a variable forage supply and we have added costs to our operation.

Because we have a variable forage supply, we need to plan our farms and ranches to have variable animal demand. More specifically, we need to manage for a variable stocking rate. We need to create more animal demand when forage supply is high and we need to be able to reduce demand when supply is reduced. And, we need to know how to monitor forage supply so we know when those peaks and valleys are likely to occur.

The narrow view of stocking rate is it is only about the number of animals on the pasture. The way we really need to view stocking rate is as animal demand for forage. Think about the changing forage demand as a cow goes through her stages of production. Demand is highest at peak lactation and lowest post-weaning. We've already learned change in demand may be anywhere from 30 to 80% for beef cows depending on milking ability. It can more than double for dairy cows or ewes with twins or triplets.

Think about this in the context of a herd of 100 cows. During the dry, pregnant low-demand time frame, we can say we have 100 cows on the pasture. After they begin to calve and their demand increases, they are equivalent to having 130 to 180 of the dry, pregnant version of themselves. There is an automatic variable stocking rate built into every herd of breeding/lactating females. In some forage environments, that built-in variation may be all the management flexibility you need to bring forage supply and animal demand into balance. Nothing more than shifting calving season to the right time of the year could eliminate the need for making and feeding hay.

For a lot of us, it's not that simple. The summer forage peak is too high and the winter valley is too low. If you have multiple livestock enterprises, getting your birthing season right is a key first step. After making that adjustment, we can start looking at adding and removing stock.

We have three basic strategies for increasing animal numbers on our farms and ranches: 1) internal increase, 2) purchasing additional animals, or 3) custom grazing. Each or these have their own set of opportunities and challenges. Every operation needs to determine what are the best options for their set of resources and market opportunities.

Internal Increase: Internal increase simply means keeping the progeny of your breeding herds beyond a traditional weaning age or weight. In a traditional hay-based, February-March calving, cow operation this makes little sense. By the time calves are weaned and winter rolls around, they weigh 600+ lbs and are expensive to carry through the winter. By the end of winter, they weigh 800-900 lbs and conventional thinking says they are too big to go back to grass. All you have done is added more winter demand for expensive harvested feed.

Sheep are a different situation from cattle. Because ewes come to peak lactation so much more quickly than do cattle and, relatively speaking, their demand is proportionally greater, a ewe flock lambing at the beginning of the grass season can add a nice spike in forage demand for the 45-60 days of peak spring growth. Just remember it takes a lot of sheep to make a difference. If we use a ratio of six ewes equivalent to a cow, it takes 300 ewes to roughly equal 50 cows. Since lambs can finish in the same growing season as they were born, there is no need to carry extra stock through the winter for finishing the following season.

Let's look at a couple of other cattle scenarios based on calving season.

A June-born calf may only weigh 350-400 lbs at the conventional October weaning season and the cow has just passed peak lactation. The cheapest way to take this calf through the winter is leaving it on the cow as she grazes stockpiled pasture. You can let the calf forward creep graze to select the highest quality stockpiled forage while leaving adequate forage for the cow. The little bit of milk she is giving is a nice protein supplement for the calf. Graziers using this system report calf gains of 3/4 to 1 lb/day. That is adequate gain to ensure full compensatory gain on spring grass. By retaining the calf all the way through winter, you now have a stocker program in place. The calves will consume the equivalent of 50-70% of a cow, depending on their weight. Combining the increased demand of the cow with the stocker phase calf, the spring-summer forage surplus may be brought into better balance.

If you plan to leave the calf on the cow over winter, closely monitor cow body condition. Body condition is the factor best indicating when the calf should be weaned. Easy fleshing cows in a good stockpile environment (e.g. fescue country) should be able to carry the calf all the way through most winters. In less favorable environments, it may be necessary to wean the calves before spring arrives. In this case the highest quality stockpile or winter annual pastures should be allocated to the weaned calves. Cow BCS can be allowed to slip into the low 4 range because cows will still have at least a month of good spring grass before calving. Letting cows fall below BCS 4 can jeopardize rebreeding opportunities. Regardless of calving season, BCS 5 is still the optimal BCS for calving.

Fall calving also provides a nice stocker opportunity for the following season. Fall calving requires higher quality forage through the first few months of winter than required by the

June calving cow discussed above. Peak lactation and breeding have already occurred with the June calving cow so her needs are on the decline. Yes, the calf's requirement is steadily increasing, but it's less than half the size of a cow. In the fall calving situation we expect peak lactation and rebreeding to occur with the onset of winter. Most fall calving operations are located in areas with milder winter climates, but some folks still try to do fall calving in pretty tough environments.

If you have real winter, remember the cow's energy requirement is going to increase by 1% for every degree below the critical thermal level. The incremental weather increase is what might put fall-calving cows over the edge when it comes to maintaining adequate body condition.

While fall calving is the best fit for most of the tall fescue belt, we have seen fescue use steadily creeping north over the last few decades. The ability to stockpile fescue in the northern Corn Belt is certainly there, but the weather may be too severe for fall calving. Stockpiled fescue or any other pasture can still be a valuable resource for dry, pregnant cows or ewes, just not for peak lactation.

Purchasing additional animals: The one real advantage of buying additional stock for consuming the spring-summer forage surplus is you can get exactly what you want, when you want them. If you've been doing a good job of conducting your pasture inventory every two weeks and your calculations say you need 47 extra cows from May 3 through August 6, you can go out and get them when you need them and have them sold when you need them to be gone. Well, maybe that only works in the perfect world.

When we lived in Missouri, there were a dozen sale barns within 50 miles of our farm. There was a sale somewhere on almost any day of the week, so you could conceivably buy or sell animals as needed. The upside was you could really manage stocking rate if you wanted to. The downside, of course, is falling into the buy high-sell low routine; transportation and commission costs for shuffling animals around; the time required to do all the transactions; and the risk of bringing unhealthy stock to your place. Most of these downsides are manageable and in some cases are more perception than reality.

Where we live now in Idaho, the nearest livestock auction is 135 miles away and there are only three or four sale opportunities within 200 miles so the buy-sell opportunity here is very different. Different situations call for different strategies. I don't think short term buying and selling is a viable strategy for balancing stocking rate in most of the Western states.

If you can buy stock directly off another farm or ranch without going through the sale barn, that is the preferred avenue. You may have better assurance of the stock being healthy and both parties can avoid commission and yardage costs. The challenge here is finding someone who wants to part with some stock during what is likely to also be their peak pasture system. If you buy from another region where prices are low due to drought or some other disaster, transportation costs become a very significant consideration.

Many readers are likely to be aware of the Bud Williams "Sell-Buy" marketing philosophy and wonder why I am saying "Buy-Sell." The Bud Williams program works well when you are continually bringing stock into an operation and selling groups on an ongoing basis. In that scenario, every time you sell a group of stock you buy back another group under the same market conditions. What we are considering here is buying one group of animals at the beginning of the peak pasture growth season and selling the same animals when the peak has passed. It would then be 7-9 months before another group was purchased for the next season's growth peak. In this case, you cannot sell and buy on the same market, so you are locked into a buy-sell marketing sequence.

Let's look at each of the challenges we've identified to buying and selling livestock to manage forage supply and how we might deal with them.

There are several ways to avoid the buy high-sell low syndrome. The first is just being judicious in what you buy. What commands the highest price at the sale barn? In most of the USA today, it's the slickest, fattest, blackest steers on the sale bill. One of the best ways to avoid the buy high-sell low problem is don't buy the highest priced animals in the first place. You have very little room to improve these animals while they are on your property, so the potential profit margin on them is relatively small.

Think about ugly animals instead. In my experience, the greatest margin opportunities come from buying ugly animals in small groups and consolidating them into a larger group. If every cow in a pot load is equally ugly, they don't look half bad. The same thing can be said for buying steers or heifers. A pot load of red and white spotted steers will command a 10-20% higher price than the same set of animals sold in sets of two or three.

Thin cows coming out of winter can gain 4-5 lbs/hd/ day. The cull cow is the only animal in the beef complex that gains in value per pound as she increases weight and gets fatter. If you determine you only need extra animals on the place for 60-90 days, the thin cow is the ideal animal for your operation.

Moving small groups of animals around is always relatively expensive. If your operation is large enough to deal in pot loads of cattle (45-50,000 lb lots), that is the unit you should plan to transport as it is the most cost efficient. If your operation needs fewer animals and you deal in stock trailer loads, the first thing to realize is you cannot justify owning a stock trailer and the big pickup needed to pull it around. Hiring your hauling will be much cheaper in the long run.

You can waste a lot of time hanging around sale barns trying to find just the right set of stock to meet your needs. Do not hesitate to use an order buyer or cattle broker. As long as you set the parameters for what type of cattle you want to purchase and the price range, order buyers and brokers will do as good or better job of getting you the stock you need. They can save you hours of wasted time and in the end they often save you money.

My experience buying both sheep and cattle at livestock auctions has generally been positive. A big part of the sickness occurring in so-called sale barn cattle is a result of how we handle them when we get them home. While stock can be exposed to various germs at the sale barn, not many actually get sick there. Most get sick because of the stress we place on them when we get them home. If you get them promptly out on good quality pasture and don't apply working pressure to them, they are usually fine. When dealing with sheep, unload them on your place through a foot bath to prevent bringing foot rot on your property.

Custom grazing: Custom grazing is another very good way to increase your peak growth season stocking rate without having to spend a lot of money. Most people automatically think beef steers when custom grazing is mentioned, but there are many other custom grazing opportunities that may better suit your needs. The most important aspects of custom grazing are developing a good working relationship with the cattle owner or broker and having a written contract.

It is far better to work with the same owner/broker year after year so the person you are working with knows what your goals are and why you need the type and numbers of livestock in a particular time frame. If they know what you're trying to accomplish and that you're a reliable grazier, they will find a way to work with you. The written contract ensures both parties know what their respective rights and obligations are. The contract doesn't have to be complex. We used a simple one page contract, but it covered the necessary points and eliminated any possible "I said, you said" arguments.

Beef cow-calf was the primary enterprise on our farm in Missouri and our goal was to develop a make-no-hay-feed-no-hay operation. Our goal with custom grazing was to take all the excess spring-summer forage that traditionally went into hay for winter cow feed and put it into a directly marketable animal product in that season. We needed the contract animals off our pastures by early August so we could stockpile feed there for winter cow feed. We were basically using the same acres as before to provide winter cow feed, we were just changing the pattern of use.

At different times we custom grazed beef steers, re-placement heifers, cow-calf pairs, dry pregnant fall-calving cows, and even horses. The steers were done on a weight gain basis while all the others were done on per head/day basis. As a general rule, anything you can do on a set per head basis is usually more profitable and less risky.

For us, the best fit was the dry, pregnant, fall-calving cow. They fit our window of need ideally. They had big appe-tites but lower nutrient requirements. We could make them eat almost anything and they still gained or at least maintained themselves. We knew on the day they arrived what our pay-check for that enterprise was going to be. That in itself is worth a lot.

Yes, you can make good money custom grazing steers, but you have to do everything right and you need to make sure they leave your pastures while they're still gaining at a pretty good clip. If you keep stockers on your pasture until their rate of gain has slowed to half of what it was at the beginning of the season, you will lose most of the money you earned in the first couple months they were there.

Within the context of our goals, what we found most challenging about custom grazing was getting the extra animals off our pastures when they needed to be gone to allow those pastures to stockpile for the cow's winter feed. The first couple years of custom grazing we did not have written contracts allowing the cattle owner to keep stalling on when the cattle were removed. When the cattle ended up staying well into September rather than leaving in early August, we lost the opportunity to stockpile feed and had to rely on hay for winter feed.

The single most important thing having a written contract did for us was setting a date by which the contract cattle had to be off our property. If they weren't gone, we had the right to charge yardage in addition to the regular fees. Just a suggestion, set the penalty yardage high enough so the cattle owner isn't even tempted to leave the stock there beyond the contracted period.

Chapter 9
Designing your winter forage systems

For most of the northern half of the USA and Canada, winter is a time of year when there is no pasture growth. In the southern USA, winter may be more a time of just slower growth or, as in the Deep South, the season when winter annual forages thrive. The duration and severity of winter changes as we move from South to North. Your winter grazing plan needs to match the conditions where you live and utilize the resources you have available.

Planning for winter grazing is largely a matter of balancing supply and demand. In the previous chapter we introduced the idea of stocking the farm or ranch based on its winter grazing capacity, rather than the summer carrying capacity. This is a guideline pertinent to an operation where a base herd of cows, ewes, or other breeding females forms the centerpiece enterprise. Adding or removing other stock provides the seasonal adjustment to demand.

Stocking for winter grazing capacity requires a planning strategy fundamentally different from conventional thinking. The conventional rancher who decides to start being unconventional approaches the idea of winter grazing from this perspective: "Okay, I have 500 cows. How do I provide winter grazing for 500 cows?" This rancher might do a lot of planning, come up with some very good ideas, but still come up short on winter forage supply.

Why? Because the 500 cows are based on summer grazing capacity. We can ask, why do you have 500 cows? A common answer is that's how many it takes to make a living. That might be how many it takes to make a living in conventional ranching. If the goal is to earn $50,000 after all expenses, that means 500 cows with a net return of $100 per cow.

What if the profit margin were $200 per cow? If the goal remains earning $50,000 then it only takes 250 cows to earn the same salary. Wouldn't it be a lot easier to provide winter grazing for 250 cows rather than 500? The quickest way for most cow-calf operations to add $100 or more net margin per cow is to eliminate hay feeding.

The unconventional rancher who is already a crackpot grazier is going to approach the same question of winter grazing from the opposite direction by first assessing how many total stock days of winter grazing are likely to be available. By knowing the duration of winter, the daily requirement per cow, the rancher will then calculate how many cows can be carried through the typical winter. He/she will already have a contingency plan in place for dealing with the atypical winter. This is what we mean by stocking the farm or ranch based on winter grazing capacity.

The conventional rancher incurs costs by making hay out of the inevitable spring forage surplus wile the unconventional rancher earns income by grazing the spring surplus forage. The conventional ranching community automatically assumes the outfit running 500 cows is the better operation, but the reality is the unconventional operation is likely to be far more profitable. Stocking based on winter carrying capacity is a major paradigm shift and one of the foundations of profitable livestock production.

We considered the winter non-growing season for perennial pastures in north Missouri to be from mid-November through March. By using winter annuals the winter non-growing period could be shortened to December through early March. There are always those seasons when autumn lingered much later or spring came early and the opposite when winter came early and stayed late. Using these criteria, we could say our winter lasted from 100 to 150 days. We typically used 130 days of non-growth as our planning window for winter grazing. That is the time period in which we expected no new pasture growth so all grazing had to come from forage stockpiled during the growing season or we needed to use winter annuals to extend the growing season. A 130-day no-growth period meant we needed to plan to stockpile about 1/3rd of the farm through late summer and early autumn.

If you're planning on using perennial pastures in a high rainfall or irrigated environment, simply taking the percentage of non-growing days in the year and applying that same percentage to the total pasture acres available gives a good starting point for planning how much to stockpile. For example if you have 120 days of winter dormancy, that is about 33% of the year on dormant feed. Plan to stockpile 1/3rd of your pasture acres. No, you're not going to come out exactly right the first year you try it, but you will begin to learn where you need to make adjustments.

In central Idaho the growing season for rangeland is typically less than 100 days and frequently no more than 60. Using the same logic in a semi-arid rangeland environment suggests a year-around grazing operation relying strictly on rangeland would need to stockpile a minimum of 75% of the total acres. The rangeland reality, however, is an even higher percentage of the land base needs to be stockpiled because of the inevitable weather-induced deterioration that will occur when rangeland grazing is deferred for 9-10 months. Wildlife will also take their share of the forage resource. The greater the wildlife population, the more area that needs to be stockpiled.

How much weather and wildlife adjustment should be made in planning stockpiled rangeland acres? Expect a 3 to 5% reduction in forage volume per month forage use is deferred. If growth ended around August 1 and the pasture isn't grazed until April, the expected April carrying capacity for a particular pasture would be about 25-40% less than what the same pasture could have carried had it been used in August. Different range sites have more or less deterioration depending on species composition, exposure to wind and snow, and wildlife pressure. You may need to adjust the projected deterioration rate upwards or downwards based on actual conditions. The best way to determine what you need to do is by diligently maintaining an ongoing pasture inventory,

The growing season on irrigated pasture here in Idaho may range from 100 to 150 days, depending on season and water availability. If these irrigated pastures are the only forage resource in the plan, we would need to have stockpiled at least 2/3rds of the acres. How could we accomplish that when we needed to be grazing those same acres through the growing season? This is why we need to have a well designed plan to accomplish year-around grazing. It won't happen by accident.

Knowing the relative duration of the growing and non-growing seasons is one of the first steps in the planning process. In the three examples described above, we had three distinctly different relationships between the potential growing season and the expected dormant season. They call for different strategies to accomplish the goal of year-around grazing.

In Chapter 5 we identified five possible sources of winter forage your livestock might be able to utilize. Those were: 1) Crop residues, 2) Cool-season perennial pastures, 3) Warm-season perennial pastures, 4) Native rangeland, and 5) Annual pastures. You need to determine which of these are viable options for your operation. Remember, these do not necessarily have to be on your property and crop residues and annual pastures do not have to be from crops you yourself are producing. We also showed how to make a chart of when these resources might be available for grazing.

Look at your own situation and make a bar chart of which of these resources you may have available for your operation. If there are one or more winter months not spanned by one of your bars, it is time for some brainstorming. Bringing in other family members, hired help, or crackpot-grazier neighbors might bring to the table an option you hadn't thought about. Lack of creativity is why most of your neighbors are feeding hay. Don't be limited by, "This is how we've always done it." or "That just won't work here." Push the envelope!

If you live in the rangeland environment, what could switching a BLM lease from summer grazing to winter grazing do for your operation? If you live next door to farmers, what about renting their crop land for winter annual pasture over sown in standing corn? Is there a nature reserve that could be grazed in late summer while your home pastures are stockpiling? Each of these is something I have seen done by farmers and ranchers who were willing to think outside of the box.

The next step, after determining what forage opportunities exist, is calculating how many total stock days might be available from each source and when they are available. Total stock days is the combination of acres available and the expected yield per acre. If you have 100 acres of stockpiled tall fescue available and the expected yield is 90 CDA (or whatever unit of measure you choose to use), the total stock days available from that resource is 9000 cow-days (100 acres X 90 CDA).

If you have 50 cows, there are potentially 180 calendar days of grazing available. That is more than the entire winter in most places. If you have 100 cows, there are 90 calendar days of grazing. If there are 500 cows, there are only 18 days of grazing available from that field. This is how simple the calculations are. The most difficult part of the process is accurately estimating the yield of the pasture. Chapter 6 discussed how to estimate yield in either pounds of forage dry matter per acre or cow-days/acre (or whatever unit of measure you choose to use).

Make the calculation for total stock days for each of the grazing resources you have available on a month by month basis. Ideally you want the number of available days to be increasing thru the winter months. If the line is flat or decreasing, you have no margin to account for increased weather and wildlife losses or the gradually increasing animal demand as they approach the end of pregnancy.

Planning the appropriate sequence of use for the different winter forage resources can make a huge difference in the success of winter grazing. Pasture composition, landscape location, stock water availability, along with other factors unique to your operation determine the optimal time of use for different pastures. The better you understand differences in rates of forage yield and quality deterioration, wildlife preferences, weather patterns, and animal behavior, the more effectively you can manage your winter forage supply. All of these factors will be addressed in subsequent chapters.

Thus far, we have primarily considered just maintaining a herd of spring birthing-summer breeding females over the winter. This is the simplest form of winter grazing as the stock in question have relatively low demand for both forge quantity and quality. When we consider keeping late spring or early summer-born calves on the cows over winter, the yield and nutrient demand moves up a step. With fall calving cows it goes up another notch. Taking growing stock through the winter at an acceptable rate of gain demands even higher quality forage. As the needs of the stock you plan to take through winter increase, your level of management will also need to increase.

The chart we made for forage availability through each month of winter now needs to have a forage quality factor added to it. Of the pasture available, how much is growing or lactation quality feed and how much is just maintenance feed? Not only do we need to plan the sequence of pasture use, but also the sequence of animal use.

Crop residues are rarely growing quality feed and should be used primarily for dry, pregnant animals or late lactation beef cows. Stockpiled cool- or warm-season pastures may contain some growing quality feed depending upon how they were grown. Stockpiled native range can support some winter stock growth, but generally requires some supplementation. Native sub-irrigated meadows may contain a relatively high percentage of growing quality feed. Winter annual forage is the best option for young, growing animals in a winter grazing program.

Every pasture, whether it's actively growing or stock-piled dormant season pasture, contains a range in quality. Stems are low quality while leaves are higher quality. A pasture that is entirely leaf growth has its highest nutritive value in the upper part of the canopy while the lower part of the canopy is less digestible. Thus, even stockpiled pasture may contain some growing quality feed. The majority of the forage is likely to be just maintenance feed. If you have both growing stock and dry, pregnant females, the pasture can be used sequentially with the growing stock allowed to harvest the higher quality forage and the dry, pregnant females left with the remainder.

The question becomes how much of the stockpile is growing feed and how much is maintenance feed. The way this can be determined in a research setting is by clipping off entire plants and then slicing them into one-inch increments and doing a quality analysis on each one-inch strata. Tedious and expensive, but pretty definitive. You are not going to want to do this, but the work that has been done can give us some guidelines.

In the northern half of the USA, stockpiled cool-season perennial pasture will rarely have more than 20% of its volume as growing quality forage. As we move south, the percentage can increase to upwards of 50% in certain situations, especially in an endophyte-free tall fescue with legume mixture. Using the 20% as a planning figure, we can easily calculate how many 500 lb weaned calves we might be able to carry through the winter in conjunction with the base cow herd. Chapter 21 deals with budgeting winter forage and will explore these calculations in greater depth.

The situation with stockpiled warm-season forages is quite a bit different. Stockpiled bermudagrass or bahia grass does not contain much growing quality feed. It is much more likely to be maintenance quality feed for dry, pregnant females, so your planning strategies will be different in regions dominated by warm-season forages.

Winter annual forages are typically much higher in both protein and digestible energy than are stockpiled perennial pastures. That is mostly attributable to the fact the winter annuals grow at much lower temperatures than most perennials. Even in northern latitudes, winter annual forage can provide an acceptable rate of gain for growing stock through much of the winter. There is a great deal of variability in how well annuals weather the rigors of winter, and making the right species selections can have a tremendous impact on your success.

Look at all the resources you have available and then sort out how they best fit your wintering strategies. Always remember it is easier to change your livestock than your forage base.

Chapter 10
Understanding animal needs and forage quality

The daily intake of feed by livestock, or humans as far as that goes, is partitioned for different bodily functions and activities. Requirements can basically be split into maintenance and production functions. Production functions include such things as lactation, growth, and work.

Mature cows or ewes can be thought of as having maintenance and lactation as their primary needs, while a first calf heifer has maintenance, lactation, and growth. A steer being finished for beef has maintenance and growth functions. A mature draft horse has maintenance and work functions to be met.

Maintenance is always the first need to be met. This is the basal metabolism that keeps us standing up and breathing. For domestic ruminants the maintenance requirement ranges from an intake level of about 1.2% to 1.9% of bodyweight. Sheep have a lower basal maintenance requirement than do cattle. High milk producing breeds or individuals have higher maintenance requirements than similar sized animals with lower lactation potential even when not lactating.

Weather stress can increase maintenance requirements. Both cattle and sheep have much lower thresholds for feeling cold than do humans, particularly when they are dry. Think in terms of leather and wool. The lower end of the thermal comfort zone for a beef cow with full winter hair coat is about 18

degrees F. Below this temperature, the daily maintenance
energy requirement increases at a rate of about 1% per degree
F. This value is for English and Continental breeds (Bos
taurus). The Bos indicus breeds have a higher temperature
threshold and have even higher additional maintenance demand
in cold weather.

It may come as surprise to many readers that the lower
critical temperature level for sheep is actually higher than it is
for cattle. Sheep, even with 4" fleece cover, begin to feel cold
at about 26 degrees F. The difference isn't because leather is a
better insulator than wool. It's due to the amount of skin surface area relative to body mass. Sheep lose more heat per
pound of bodyweight than do cattle. It's a good thing sheep
have a lower basal metabolic rate to allow them to get through
tough winters on limited forage resources! Newly shorn sheep
or goats have much higher maintenance requirements than do
their counterparts with full fleece or hair coat.

The protein requirement for maintenance is between 6%
and 8% depending on animal species and milking ability.
Energy requirement can be expressed as total digestible nutrients (TDN) and is usually between about 48 and 52%. Remember only energy demand increases due to cold weather, not
protein.

As production levels increase, both energy and protein
requirements go up. The higher the expected level of performance, the higher both energy and protein go. High milk
producing beef cows have a peak protein requirement around
11%, while dairy cows and sheep may peak at 14-16%. As
expected average daily gain (ADG) increases, so do energy and
protein requirements.

There are two basic patterns describing the nutritional
needs of your livestock. In breeding herds there is a very
consistent and predictable cyclical pattern to their needs, while
in growing stock there is a steadily increasing demand for
forage as the animals gain weight. Understanding these patterns
of needs is critical to planning successful winter grazing.

Nutritional demand cycle for breeding herds:

The basic pattern of nutritional demand in breeding herds is a gradually increasing need for both quantity and quality of forage beginning in late gestation, a rapid increase from birthing to peak lactation, then a slow decline through later stages of lactation, and finally the lowest demand in terms of both quantity and quality occurring after weaning. The extent and pace of the increasing demand is largely driven by the animal species and the milk production potential of the dam. There are also weather and other stress factors that can affect nutrient requirements.

Cattle, bison, sheep, goats, elk, deer, and other ruminants all have this pattern. With the larger ruminants like cattle, bison, and elk, the cycle occurs over a 12-month period. Cattle have a little over a nine months gestation period so there are fewer than three months from calving to rebreeding. Typically we expect a cow to rebreed within 83 days of calving. These three months from calving to rebreeding are the most critical nutritional period for a beef or dairy cow. This is when she will eat the greatest volume of feed and it is also when the protein and energy requirements are the highest. Ideally we would like this peak demand to occur when we have an abundance of cheap feed.

Cows come to peak lactation in about 45-75 days, depending on breed and milking potential. Cows with higher milking potential require both more feed per unit of body weight as well as more nutrient-dense feeds. Even when they are not lactating, high milk producing females have higher maintenance requirements than their more moderate milking cousins. This is an important consideration when selecting cows for a year-around grazing operation.

By maintaining a 12-month calving interval with cattle, the peaks and valleys will always fall at the same time of the year. Knowing when the peaks and valleys of animal demand occur and planning them to coincide with the peaks and the valleys of forage supply is a fundamental principle with year-around grazing. The greater the milking potential of the cow, the higher her peak demand is going to be relative to her lowest demand level. For year-around grazing, you want to stay away from high milk producing cows.

What is high milk producing? Any female sired by a bull with above breed average EPD for milk production will probably fall into the high milking classification. For many breeds, you might want to actually be looking for negative milk EPDs to bring things back into balance. Remember, milking EPDs describe a production trait, not maternal characteristics of a cow.

The increased nutrient demand of a high milk producing cow may be 80% or more compared to her demand as a dry pregnant cow. An old fashioned "range cow," the kind that stays fat on sagebrush and buffalo grass, may only increase 25-30% with peak lactation. These differences account for the huge hay consumption requirements of the high milk producer. Remember, even at maintenance, the high milk producer has a higher nutritional demand. Which cow do you suppose is easier to keep in a year-around grazing system? Which is likely to be more profitable?

Selecting the right type of cow for your resource base is an important part of the year-around grazing equation. There is a great deal of variance in the yield and quality potential of the five different winter grazing options covered in this book. The greater quality potential in your winter grazing resources, the more milking ability you can tolerate in your cow herd. You still don't want the top milking dams in any breed. This is a good place to remember the old maxim of moderation in all things.

With smaller ruminants like sheep and goats, the duration of the gestation-lactation-weaning cycle is much shorter. Gestation length is a little less than five months for both sheep and goats. In naturalized herds, they tend to breed in late autumn or early winter and give birth in mid-Spring. They come to peak lactation very quickly, in just two to three weeks compared to the seven to ten weeks for cattle. While cattle may stay near peak lactation for a month or more, ewe lactation generally begins to fall off within a couple weeks of reaching their peak. Nannies from dairy breeds will maintain high lactation for longer periods of time.

If ewes or nannies only lamb or kid once a year, the period of time from when the lambs or kids are weaned until the dam is rebred may be several months, depending on age at weaning. This means the small ruminants have a longer period of time when their nutritional requirements are very low.

As with cows, ewes and nannies breeding and birthing only once a year have a very predictable calendar for when their requirements are going to change.

What about the accelerated lambing programs designed to produce three lamb crops in two years? In a system like this, the calendar timing of when gestation, lambing, and weaning occur change with each eight month production cycle. While it can be accomplished in many environments, the challenges for year around grazing increase exponentially with accelerated lambing programs.

Nutritional pattern for growing stock:

What would it mean if your yearling steer or feeder lamb had the same nutritional requirement day after day for a month?

If you formulated a ration or set a pasture allocation based on their weight on the first day of the month and then managed them identically for the next 29 days, they would either be gaining no weight or gaining at a steadily decreasing rate. Why? Because on each day more of the feed consumed that day is going towards maintenance. As a general guideline, the feed requirement of a growing animal increases every day. So the question becomes, what do you want the animal to be gaining through the winter months?

One way to answer that question may be whatever you can afford to have the animal gain. Year-around grazing is largely about reducing production costs so you might think in terms of getting the stock through the winter at the lowest possible cost.

What happens if the animals gain no weight or even lose weight over winter? You don't have much invested in them, but what have you lost?

Time is one obvious answer. If your winter period was four months and the stock gained very little or no weight, they have to make that time up in the growing season. If it puts their target finish weight out of reach in the next season, you're looking at another winter of no gain. Eventually, the high cost of time will catch up with you.

What about compensatory gain? Can't the stock make it up with explosive growth in the spring?

There is a minimal growth rate that must be maintained for an animal to finish at its genetic potential weight. If the animal doesn't achieve that rate of gain for a period exceeding 30-60 days, it will not reach its genetic potential. It is permanently stunted.

Most research suggests that rate of gain for cattle is about 0.7 lb/day. For low-cost, rough wintering this should still be the target rate of gain if your plan calls for full compensatory gain to reach a target finish weight by a certain date.

What you need to know about forage quality:

Forage quality is mostly about having digestible energy and protein available in the plant in the appropriate balance. Yes, minerals and vitamins are also important, but protein and energy are the biggest consideration.

Energy is derived largely from digestible fiber by ruminants while protein is usually associated with leaf material. In green plants there may also be substantial energy available in the soluble cell contents that contain the direct sugar products of photosynthesis.

Cell solubles are much less important sources of energy in dormant plants commonly being grazed in the winter months. Without adequate protein in the rumen, fiber cannot be digested to provide energy to the animal.

Plants differ in their energy and protein potentials depending on broad plant classifications as well as individual species. Here are some of the general guidelines.

■ Annuals are higher in both protein and energy than are perennial plants.

■ Cool-season plants are generally higher in protein than are warm-season plants.

■ Warm-season plants often have higher energy potential than cool-season plants.

■ Legumes are higher in protein than grasses.

■ Growing plants are higher in protein and energy than are dormant plants.

Within any plant group, maturity is the number one factor affecting fiber digestibility and protein availability. As plants mature, the digestibility of fiber decreases and the indigestible components of the plant increase. Protein can become bound to the indigestible fiber and become unavailable even to ruminants.

When dealing with stockpiled perennial pastures, quality will always be lower than for the same species at a comparable maturity stage during the growing season.

Some winter annual forages can maintain quite high levels of digestibility and available protein even under snow or ice. Examples are annual ryegrass, wheat, barley, and triticale. Other winter annual forages can turn to mush in the same situations with most brassicas being the best example. Forage quality is best preserved during the winter with a layer of dry snow covering most of the pasture. Repeated freezing and thawing or rainy conditions in the winter are most detrimental to forage quality.

Collecting forage samples through the winter to monitor the nutrient content of your winter forage supply is a good idea.

The forage sample most accurately representing the animal's true diet is one pulled from the rumen after a few hours of grazing. Most of us don't have cows with access ports in their sides, but we did a lot of work with cannulated animals while I was at FSRC. By collecting direct plant samples from the pasture at the same time we were collecting rumen samples, we were able to determine how much better a cow is at grazing than what a graduate student can do.

There are two basic ways of collecting forage samples from a pasture. One is to clip off the entire plant, much like hay would be, and send that in for analysis. We'll call this a whole plant sample. The second method is to watch grazing stock and try to pluck the same plants and plant parts the stock are grazing. We'll call this the plucked sample.

If you are taking forage samples from your pastures, keep these relationships in mind. A cow will typically select a diet that is 2-4% higher in protein and 5-8% higher digestibility than what a lab analysis on a whole plant sample shows. She will also be able to select 1-2% higher protein and 3-5% higher digestibility than a human trying to do his/her very best job of simulating what the cow is taking. Cows are very good at being cows. Sheep can select a diet that is even a little bit higher in protein and digestibility than what a cow does.

The bottom line when you're looking at lab analysis of pasture samples, remember the animal is doing better than what is shown on the data sheet.

Plan supplementation for what the animal is likely taking by using the above adjustments for what the lab test shows. It can save a lot of money and rarely results in a wreck. This does not mean you can plan your entire wintering program on a single sample taken in November. If you are going to use forage sampling, go all the way and use it effectively.

In the first few years of implementing a winter grazing program, I suggest taking bi-weekly samples so you can monitor the rate of change in pasture nutrient levels.

Once you have a good baseline understanding of what is

going on, you might go to monthly samples. Eventually you may just rely on your eyeball and close monitoring of stock condition and behavior.

Behavior or apparent contentment will begin to change long before you pick up on declining body condition, so watch what your animals are doing as well as how they are doing.

Chapter 11
Making the transition from hay feeding to grazing

How long did it take farmers and ranchers to get into the hay feeding paradigm? Generations.

How long have most farmers and ranchers been in the hay feeding paradigm? Most of their lives.

Change rarely comes easy for most people. If your livestock operation has been relying on hay as winter feed for a number of years, do not expect to make the transition to year-around grazing in a single season. While I have seen a few operations make the change in a single year, it has usually been because they were feeding very little hay to begin with, were grossly understocked, or it happened almost entirely by accident.

If your farm or ranch is operating near carrying capacity and utilizing hay for several months, the infrastructure for year around grazing is probably not in place, your livestock may not know how to cope without hay, and you may not have the management skills or experience to make it happen. I have already emphasized the need for advanced planning several times in previous chapters. You should develop a plan, with contingencies, for making the transition from hay feeding to winter grazing.

How long should it take to make the full transition? I think it is a reasonable goal to plan to gain a month of grazing each year through changes in management. If you are feeding hay only for a month on average, you can probably make the change in a single year. If you are feeding for two months, plan to make the change over two seasons. If feeding for three months, plan for three years and so forth. For a lot of major farm and ranch transitions, I help develop three to five year management plans for my clients. Moving to year-around grazing falls right into this category. I think five years should be the upper limit to making the change to year-around grazing. If it takes longer than five years, the business may not survive.

A transition period means there will be a time period when hay continues to get fed. One of the first questions that arises is should you continue to make hay or should you buy the necessary hay? Based on relative cost of production and market value of hay, it is really more important to quit making hay than it is to quit feeding it. In the eastern half of the USA, hay can almost always be bought for less than average cost of production. This is especially true for smaller operations.

There are occasional years when hay prices surge due to drought or other natural disasters, but year in and year out since the mid 1980s, hay value has been less than production costs. You have to ask how can this possibly be? The answers lie in off-farm jobs, other enterprises subsidizing hay production, and the simple fact that it takes a whole generation of farmers to fully adjust to the changing economic climate.

Buying hay also brings valuable soil nutrients and organic matter onto your pastures rather than removing them. There was a time in 2008, the year of soaring input costs, when just the N-P-K content of a ton of hay was greater than the fair market value of the hay. Contrary to conventional thinking, you probably have greater control over hay quality by buying it than you do putting up your own. You can control what you purchase, but you can't control the weather during haying season on your farm or ranch.

If you decide to continue to produce some hay on your property during the transition, the next question is whether to put the hay up yourself or have someone else do it. If you're a small operator, equipment ownership is likely to already be one of your disproportionate expenses. As you reduce the tonnage of hay you are harvesting, the equipment cost per ton increases and hay making becomes even more costly per ton. No matter how much you love your equipment, it probably makes sense to sell it and hire a custom operator.

Custom operators often get a bad rap for not being at your place on time, making lousy hay, and a whole array of other complaints. I spent several years on the receiving end of those complaints. Through most of my high school years and all of my undergraduate college years, I worked with my dad and brothers in a custom hay business. We ran a large round baler (one of the first in Illinois), two square balers, and the accompanying mowers, conditioners, and rakes.

What I found were the people who complained the most about custom operators were the ones who called you the day after the hay was ready to be mowed or raked and wanted you there right then. No forward planning and no contingency plan. Most of our regular customers were lined up several weeks in advance and we knew when we would be there. We planned extra days in the schedule for the inevitable rain delays and equipment breakdowns. We made hay for those people year after year and they were generally satisfied customers.

The bottom line on custom operators is they need forward notification and planning and you must realize they also have schedules to maintain. If you work cooperatively with a good operator, you will most likely be satisfied.

Once you decide to get out of the hay making business, you can sell your equipment. Do not park it behind the barn or fence row thinking maybe someday you will need it again. All it does there is lose value and maintain the temptation to fall back into your old habits. Hay making is an addiction. Sell the stuff and use the money for developing your grazing infrastructure. I have seen many operations where their entire fence and stock water developments were paid for by selling some farming equipment.

You should plan to maintain a hay reserve while you are making the transition. How big it needs to be depends on your level of risk aversion, how readily available is hay in your neighborhood, frequency and duration of certain inclement weather events, your storage capacity, and the price of hay among other things.

Begin by considering how long of periods of deep snow, ice storms, or excessive mud your area typically experience. A lot of farmers and ranchers feed hay for four months because there is one month each winter when grazing conditions are challenging. For the rest of the time, grazing might be fine, providing something was there to be grazed. Plan your hay reserve to cover the duration of those periods.

In the West where hay can be left setting out in the open and experience little or no deterioration, storage capacity is not an issue. In the wetter parts of the country, ground level and overhead protection are important parts of preserving hay quality, especially if you plan on storing hay for more than a year.

Any hay storage site, whether open air or under cover, requires a raised pad with good drainage away from the storage area. For tightly wrapped large round bales, this may be all the protection you need, while loose wrapped round bales and square bales (large or small) require overhead protection. That may mean under roof or just covered with durable tarps.

I recommend exchanging your hay reserve on a regular basis. As each winter nears the end, if you are confident you will make it through winter without needing the hay, begin to sell some. This is when hay is usually the most expensive and can be readily sold at a good price. Replace that hay in the summer when prices are typically at their lowest.

I know several operators who have roofed hay sheds who sell half their hay reserve each year by selling out one end of the shed this year and then the other end next year. Once again, what makes them different from most farmers and ranchers? They have a plan.

We have already begun discussing planning the grazing process. Begin by identifying your opportunities for extending the grazing season, evaluate your potential carrying capacity through the winter, start making adjustments in your stock policy, and maintain an ongoing pasture inventory through the growing season as well as the dormant winter season.

The worst time to decide you're going to begin winter grazing is at the onset of winter. That is probably when you are least prepared. Planning for winter grazing begins before the first blade of green grass appears in the spring. If you start preparing from spring forward, you have a much greater chance of rapid progress and good success.

If it's already mid-autumn when you get the wild hair and decide to start moving towards year-around grazing, there is something you can do to extend the grazing season without extensive forward planning. Just begin strip grazing whatever standing forage you have left at the end of the growing season. It may be aftermath on a hay field or crop residue. Perhaps a pasture you moved off of in August has a little bit of regrowth. Get a reel of polywire and some step in posts and just start giving out a small strip at a time.

This is the best place to start learning forage allocation and strip grazing. The calves or lambs are probably already weaned. You have a dry female and dormant pasture. It is hard to screw up either one of those. Set the fence for one day of grazing based on your best guess and just go from there. If there is still plenty of forage out there after the first day, don't move them and come back tomorrow. If the cows are bawling and the ground is stripped bare when you come back tomorrow, you guessed wrong. Make an adjustment and come back the next day. This is how you will learn to calibrate your eye.

Joe Miller, a rancher from Salmon, Idaho, attended our four-day grazing academy in September of 1999. He had made no provision for winter grazing but was forced to move rapidly towards winter grazing due to a lack of hay. He went home from the school, bought some polywire, and began strip grazing his hay aftermath. With a once-a-week strip graze he increased his carrying capacity on the aftermath by 50% compared to what he had historically gotten from those fields and extended his grazing season by over a month. Once he began to realize the potential for year-around grazing, Joe moved in that direction and by 2006 was grazing year-around. Prior to 1999 he was feeding hay for five to six months every winter. Joe always says the first thing he changed was his mind.

If you do not already have the infrastructure in place to begin a sophisticated winter grazing plan, you can accomplish a lot with just a couple of reels of polywire and step-in posts. Depending on where you live, snow may provide all the stock water you need. If not, livestock usually don't need near as much water in the winter time so we can get by with less water development than required for summer grazing. We just need to make sure we can keep it available in freezing weather. Chapters 22-24 deal with fence and water for winter grazing.

The next eight chapters deal with managing the many potential forage resources for winter grazing. How you utilize pastures through the growing season obviously has a tremendous impact on what will be available for winter grazing.

I have often said the strongest argument that can be made for implementing MiG (Management-intensive Grazing) during the growing season is the ability it gives you to create winter feed. No, not because you can harvest more acres for hay. Because you change the shape of the growth curve, you can create beneficial diversity in pastures, you can rest pastures when they need to be rested, and because you have a plan.

Chapter 12
Using crop residues to extend the grazing season

For the grass farming purist, crop residues are probably the least romantic of all the winter grazing options; but for the practical grazier living among crop farmers, crop residues offer some of the lowest cost grazing days available. Crop residues are typically considered to be low quality and suitable just for maintenance of dry, pregnant females or mature bulls.

However, there is a lot of difference between rye straw and broccoli stubble, yet both can be considered crop residues. Understanding the relative merit of different residues and how they withstand the rigors of winter weather is key to successfully using crop residues as part of your winter grazing program.

There are some commonalities to consider with crop residue grazing. Almost all of them deteriorate fairly rapidly with winter weather. The ideal winter weather for using crop residues is to get cold and stay cold along with several inches of powdery snow. Freezing and thawing and periodic rains are the enemies of nutritive value of residues. Since you can't control the weather, plan to use crop residues as soon as possible in your winter grazing sequence.

Almost any kind of stockpiled pasture will maintain better yield and quality than do crop residues, so use the residue and save the stockpile for later. Many winter annual forages are continuing to grow while residues are available, so let the annuals grow and use the residues.

Residues from most crops vary in quality among their different components. For example, the residues in a corn field could be sorted out as grain, leaf, husks, cobs, and stalks. They are all out there in a mixture, but they vary in quality so livestock will sort through the mix and take the best out first.

The order the components are listed above is the quality ranking. If livestock are turned to a field and expected to be there for two to three months, they may do well for the first few weeks, maintain themselves for a few more weeks, and then start going downhill. This pattern is due to the livestock high grading the residue in order of quality. The entire mixture may well meet their needs, but stalk and cobs by themselves will not. If the field were strip-grazed in increments of no more than one week (preferably less than three days), the opportunity to selectively graze will be significantly reduced and a better balanced diet can be maintained for a longer period.

The shorter the grazing period, the less selectivity can be made among the different components of residue.

The value of strip grazing holds true for virtually all crop residues. Tools and techniques for winter strip grazing will be covered in Chapters 21 thru 24. The time period in which you plan to use crop residues needs to be consistent with the realities of your weather environment.

In the southern half of the USA where deep snow is not a threat to grazing, livestock may winter for several months on crop residues, as long as wet weather does not totally rot the residues or create an intolerable mud problem. In the northern region, weather conditions may restrict crop residue grazing to as little as a few weeks. Know the probability of your weather risks and plan accordingly.

Grazing corn stalks used to be very common throughout the Corn Belt when every farmer had a herd of beef cows and some sows. That was also when every 40, 80, or 160-acres had a fence around it. When fence-row-to-fence row farming became the norm through the Corn Belt, the fences and cattle soon disappeared. Stock ponds were bull dozed in to gain another acre or two of farming ground. Crop fields soon became a hostile environment for livestock.

Prior to the change in farming practices, it was common for cow herds to spend 60-90 days from late September or early October on stalk fields. Corn stalks still offer a huge grazing opportunity, but there are some changes to consider. When stalk grazing was at its peak from 1950 through the 1970s, typical grain harvest losses were in the 10-15% range. There were a lot of ears of corn left on the ground. Two notable changes have occurred in the intervening years.

Plant breeding has made corn stalks stand up better so there are fewer ears left on the ground. How did they do this? By increasing the lignin and indigestible fiber fraction in the plant. The result is not only fewer ears on the ground, but also lower forage value in the residual corn stalk. More efficient combines also leave less grain behind. Typical harvest losses now are only 2-4%. The combined effect of both of these factors is stalk fields are not as good a forage resource as they used to be. Whereas old-fashioned corn fields may have been okay for growing stock, modern corn stalks are basically just dry cow feed. But stalk fields are still very cheap dry cow feed.

Because most milo or grain sorghum is harvested while the vegetation is still somewhat green, these crops provide a moderate quality forage that is suitable even for stockers with moderate rate of gain targets. Depending on location, the window of opportunity to harvest sorghum stubble as a high quality feed can be narrow. Not only does frost rapidly reduce forage quality, but it can also significantly increase the risk of prussic acid poisoning from grain sorghum stubble. In the Southern Plains, there may be one to two months of good grazing opportunity, while in more northern areas, the window may only be a couple of weeks.

Small grain stubble from wheat, barley, oats, or triticale can be grazed, but are typically very low quality. This is one of those situations where weedier fields make better grazing. Very often fields are harvested in mid-summer so there is ample time for weed development in wetter environments. By themselves, straw stubbles almost always require supplementation of both protein and energy. In the northern grain belt where small grains are not harvested until late August or September and the climate is drier, basically all you will have in the field is straw. Even with supplementation, straw is usually just maintenance feed for dry cows or ewes.

An increasingly popular way to utilize small grain residue for winter grazing in Canada and on the northern USA Plains is using a chaff collector behind the combine rather than widely dispersing the fine chaff on the ground. The chaff collector dumps piles of chaff whenever the collector reaches a certain fill level.

Why is this advantageous? Chaff is usually more nutritious than straw. Loose chaff scattered over the field provides virtually no grazing opportunities as livestock do not lick it from the ground. A pile of chaff can be eaten almost like feed in a bunk with minimal waste. Just the chaff piles can provide 10 to 20 CDA of low-quality winter feed, but it's still better than just straw.

The very best crop residues are those left after vegetable crop production. Sweet corn, pumpkins, dry beans, field peas, broccoli, cauliflower, and many other veggies leave behind palatable and moderate quality residues. In the case of broccoli, cauliflower, cabbage, and other brassicas, the quality may be better than the best stockpiled perennial pastures. The problem with most of these crops is they tend to be produced in areas dedicated almost exclusively to vegetable production, and livestock are a rarity.

The farmer's objections to allowing cattle on these fields and soil types are probably legitimate. Trampling damage to raised beds or irrigation furrows can be significant with cattle. Sheep and goats provide a much better grazing alternative than cattle on vegetable residues. Using electric netting for the entire fence system assures the stock will be contained and provides an easy means of moving rapidly across a residue field. Sheep and goats have very low water requirements in the winter and can be watered using hauled water much more easily than can cattle. Sheep and goats will also do a better job of cleaning up broadleaf weeds present in the field.

The small farm with many mixed enterprises may have the best opportunity for maximizing the use of vegetable residues in their grazing programs. Lawrence and Ursula Holmes manage Cresset Community Farm, a CSA located near Loveland, Colorado. They run a combined dairy, meat and poultry, and vegetable operation. Each year a portion of the pastures are rotated into that year's vegetable gardens. After vegetable harvest, beef cattle graze the residues. When the vegetable residues are fully utilized, the garden areas are seeded to winter annual forages to provide high quality grazing for the small dairy herd. The most productive land on the farm is managed very intensively and helps provide them with extended season grazing opportunities.

If you are a crop farmer yourself, it is certainly in your best interest to utilize the forage remaining on the field after the crop has been harvested. Economists usually assign the full cost of producing grains or vegetables to the crop itself. If you use that accounting strategy, crop residue becomes essentially free grazing days. If you are losing money on the crop, you are at least salvaging part of your investment by grazing the residues. If part of the crop cost is assigned to the value of residue grazing, it is still usually less than 20 cents/cow-day.

As a grazier living among crop farmers, you may need to sell your neighbors on the idea of grazing on their crop land. Unless it is already an established practice in your area, you may meet with a number of objections. These include grazing causes soil compaction, the field will be too rough to farm in the spring, there is no fence around the field, there is no stock water available, among others.

Iowa State University and the National Soil Tilth Lab have conducted extensive studies to assess the effect of winter grazing crop fields on soil compaction and subsequent crop yields. They found no causal relationship between grazing crop fields and soil compaction, even on high clay content soils. Neither soybean nor corn yields were reduced in the year following grazing.

Benefits of grazing corn stalk fields prior to a soybean crop included reduced presence of volunteer corn and conversion of slowly degradable corn stalks into more rapidly decomposed manure. University of Nebraska has also conducted extensive work on the impact of residue grazing on subsequent soil conditions and found no incompatibilities.

Many corn fields get grazed with nothing more around them than a single strand of polywire. Cattle that are well trained to electric fence will be no problem grazing stalk fields, unless left there beyond the availability of useful residues. If you provide the fencing and oversight, the farmer need have no additional expenses for the grazing enterprise.

Stock water can be more challenging and in many situations, hauling water may be the only viable short term option. If you can arrange a longer term lease, it may be economically feasible to install a pipeline. More about fence and water in Chapter 22 and 23.

When looking at leasing crop land for winter grazing, also look for forage in the non-cropped areas. Grassed waterways, steep-back terraces, headlands, and other soil conservation areas can carry substantial amounts of stockpiled forage. In the tall fescue areas of the Midwest and South, the stockpiled fescue in the conservation areas may be all the protein supplement livestock need to effectively utilize crop residues.

The more conservation acres in a cropping unit, the greater its value for winter grazing. Just don't tell that to your farming neighbor. In rolling crop land areas, there may also be watershed lakes or ponds that can be used for stock water.

The amount of residue left in a field behind crop harvest is obviously closely tied to the yield of the crop. Higher yielding grain crops also produce more residue for potential grazing.

For planning purposes consider corn stalk fields as having 45-60 CDA potential. Small grain stubble from 30 to 90 CDA with the lower yield potential occurring in drier environments and the higher yield where annual weeds have grown up in the stubble. Vegetable residues are highly variable and can really only be assessed based on experience with that particular crop.

Chapter 13
Stockpiling cool-season perennial pastures

One of the first things to understand as you begin the move to year-around grazing is the growing season and the grazing season are two entirely different things. Stockpiling is the process of accumulating standing forage during the growing season to be grazed during periods of little or no growth. While we generally think of stockpiling as a program for winter grazing, it can also be employed in some climates for standing forage during heat or drought induced summer dormancy.

Stockpiling is not a new concept. Wild ruminants and other grazers have survived on summer accumulated forage for millennia. Many farmers and ranchers have put livestock on regrown pastures or hay fields after frost as a routine practice for generations. What is different today is we are creating planned stockpiles of better quality standing forage based on proven, practical management practices.

University research in the USA on managed stockpiling dates back into the mid-1950s. Most of these early studies were agronomic trials evaluating different forage species and fertilization programs. Kentucky and Virginia were leading states in this early research. In the 1960s, animal evaluation of stockpiled pasture began in earnest.

Virginia, Kentucky, and Missouri were leaders in early livestock research with stockpiled pastures. During the 1970s grazing stockpiled winter pasture was already a standard man-

agement practice at the University of Missouri-Forage Systems
Research Center. Virginia, Kentucky, Georgia, and Indiana all
had animal trials on stockpiled pasture during the1970s.

Through the 1980s and into the new century, more and
more states and provinces joined in the quest for extending the
winter grazing season with stockpiled pastures. What has been
learned in all these trials provides us with a basis for reliably
producing quality stockpiled pasture in most years

Properly stockpiled cool-season perennial pastures may
be utilized through the entire winter period for some classes of
livestock. Dry, pregnant cows, ewes, or nannies can be maintained throughout the winter months on stockpiled pasture as
long as it is still accessible through the snow or ice cover.
Simple availability and accessibility are more likely to limit the
usefulness of stockpile cool-season forage than its nutritive
content.

Growing high yielding and high quality stockpile
requires forward planning and management. The following is a
step-by-step guide to growing such stockpiled pasture.

Choosing the right pasture:

Basically any kind of pasture can be stockpiled but
some pastures will do much better than others. Several early
studies compared Kentucky bluegrass, orchardgrass, and tall
fescue. In every comparison, tall fescue provided the most
stockpiled forage.

Other studies evaluated smooth bromegrass, timothy,
reed canarygrass, and other lesser known species. Many of
these small plot trials did not evaluate forage quality and none
included any animal use.

I remember in my introductory forage class at University of Illinois being told by the agronomy professor teaching
the class that Reed canarygrass was a good choice for stockpiling. In later years I learned from practical experience it is one
of the very worst winter forages. The professor was looking
only at autumn growth potential and not animal acceptance or

palatability. Yes, you can grow a lot of feed in late summer with canarygrass, but nothing will eat it after frost.

When forage quality was evaluated, once again, tall fescue rose to the top of the class. For many researchers this was surprising as tall fescue was quickly developing a reputation as low quality forage among producers and county extension personnel. Then when animals were brought into the studies, we got another surprise when animals sought out and preferentially grazed stockpiled fescue after frosts after rejecting it all summer. I have seen cows in late autumn walk by orchardgrass, timothy, and smooth brome to graze isolated clumps of fescue. And when they did graze them, it was nearly to the ground.

What makes tall fescue such a good crop for stockpiling? 1) It generally produces more forage in late summer and fall than any other perennial cool-season grass. 2) Non-structural carbohydrates rapidly accumulate in the plant in response to cooler nighttime temperatures and shortening day length. 3) It endures more freezing and thawing while maintaining nutritive value better than most other grasses. 4) It forms a vigorous sod that can withstand much abuse during wet winters. 5) It generally recovers rapidly in the spring from close winter grazing.

Through the heart of the fescue belt, stockpiled tall fescue can be routinely expected to produce between 1.5 to 3 tons of dry matter as standing forage for winter use. Most other grass species will produce between a half and two tons of stockpiled forage. Soil fertility, moisture, and stand conditions determine the actual yield that will occur on your field. I have seen stockpile yields as high as five tons/acre in exceptional growing conditions and I have seen absolutely no stockpile yield in exceptional conditions going the other direction. More about risk later.

All cool-season grasses maintain a certain level of non-structural CHO (carbohydrates) throughout the growing season. These are stored CHOs that allow a plant to maintain basic maintenance respiration and regrow after all of its leaves have been removed. Without leaves, the plant has no photosynthetic factory so regrowth energy must come from stored energy.

Many people seem to think all of these CHOs are stored below ground but the reality is they accumulate above ground in the lower stem bases of most grasses. The reason livestock graze fescue so close in the fall is they are trying to get those high-energy free sugars. They aren't dumb animals.

When you handle fescue leaves, it is easy to cut your hands due to the coarseness of the leaf. There are actually little jagged edges formed by silica crystals that create this saw-like effect. The leaf is also characterized by a heavy waxy layer known as the cuticle. It is this characteristic that makes fescue so durable to freezing and thawing. Contrast the feel with orchardgrass. It has a soft leaf making it very palatable to livestock in most growth stages. Because it is a soft leaf, it also breaks down very rapidly in response to freezing and thawing and its nutritive value quickly deteriorates.

Much of the fescue belt does not experience severe winters where the ground freezes solid for months at a time. Most of the region has periodic cold snaps interspersed among reasonably mild winter conditions. The result is soils continu- ally going through freeze-thaw sequences resulting in lots of mud. Most bunch grass pastures like timothy or orchardgrass seem to have no bottom during the wet winter periods and stands can easily be destroyed. Fescue stands up to these conditions better than almost anything else available.

The last thing about fescue that makes it so good for winter pasture is you can do just about anything to it and it will come back in the spring. Sometimes you will want to damage the stand to make it easier to interseed legumes or crabgrass, but most of the time there will always be a good stand of fescue to come back.

Does this mean tall fescue is the only forage you should try to stockpile? Absolutely not. Even the most tender of forages can be stockpiled. Just plan on using it on the front side of winter and avoid grazing when soil conditions are overly wet. Over the years, I learned diverse mixtures containing as little as 30% tall fescue could stockpile almost as well as solid stands of fescue. Because tall fescue grows so well in the fall, it tends to overstory much of the rest of the pasture. The effect is basically the same as a barn. The more tender forages remain green and fresh under the fescue canopy much longer than if they were in their own monoculture.

This phenomena was really brought home to me when we started routinely seeding tall fescue, orchardgrass, timothy, smooth bromegrass, red clover, and birdsfoot trefoil as a mixture. When these pastures were stockpiled they had the typical browned-off look of stockpiled fescue when looking at them from the field edge. When we walked out into them and pulled back the frosted fescue leaves there would be soft green orchardgrass and timothy along with clover crowns below. This really contributes to the quality and palatability of the stockpile. It's even better when the pasture gets snow covered and stays that way for months. It's just like putting the forage in the freezer and storing it.

We have seen this same phenomena on stockpiled irrigated pastures in Idaho. Even in January and February, following months of cold weather with temperatures regularly dipping to double digit sub-zero readings, after a one-day strip is grazed off, the residual grass stubble is green. Not only does this tell you how protected the lower part of the canopy can be, it is also a testimony to the quality of properly stockpiled forage.

In the Western USA and Northern Plains where meadow bromegrass is grown, it provides a lot of the same characteristics that tall fescue does in the East and South. Meadow brome grows very rapidly in the late summer and fall, as long as irrigation water is available. It is extremely palatable and retains its quality after frost. We only tried growing meadow brome in one study while I was in Missouri. It was a small plot grass variety evaluation including fescue, orchardgrass, perennial ryegrass, smooth bromegrass, and one cultivar of meadow brome. We had a serious deer depredation problem there and one of the things we quickly observed was the meadow brome plots were kept grubbed to the ground all winter by the deer while adjacent plots of orchardgrass and perennial ryegrass were left nearly ungrazed.

One advantage for winter graziers in the West and on the Plains is mud is rarely a problem. Once the irrigation is shut off and the ground dries out, almost any part of the landscape can be grazed without damaging the plant community. Natural wetlands and sub-irrigated meadows are still grazing challenges and should be grazed only when the ground freezes.

There are some pastures to avoid when it comes to stockpiling for winter. From the pasture species standpoint avoid reed canarygrass and perennial ryegrass. Reed canarygrass becomes totally unpalatable after frost. Very few animals will ever eat it. Having some along a waterway for bedding and cover is fine. Just don't expect them to eat it. Perennial ryegrass gets infected with leaf diseases and rapidly melts down to nothing if you try to stockpile it. Carrying too much forage cover into winter is a good way to kill perennial ryegrass stands.

Avoid fields with lots of weeds, particularly summer annual grasses. The nutritive value of summer weeds in the winter is pretty minimal. One exception is crabgrass. Cattle will still do a good job of utilizing stockpiled crabgrass even in the dead of winter. More about that later.

Another problem with summer annual grasses in a field you are trying to stockpile is they are still actively growing at the time N fertilizer needs to be applied. The summer annuals can take up all the N and lock it up for the entire winter leaving nothing available for your cool-season grasses.

Lowland and persistently wet fields are not very well suited for winter grazing unless you have reliably cold winters that keep the ground frozen solid. They may grow a lot of stockpiled forage, but it is difficult to utilize the stockpile due to mud problems. If you are going to use wet fields, target them for grazing in the fall when they may be drier or wait until the ground is solidly frozen.

When to begin stockpiling:
Stockpiled pasture is best suited for dry, pregnant females. It will work for cows, ewes, and does of all ruminant species. With proper planning and management, production animals like fall-calving cows, replacement females, and growing stock can also utilize stockpiled pasture for some or all of their winter nutrition.

The type of livestock you will be wintering helps determine when to begin growing your stockpile. Longer stockpile periods result in high yields but lower quality forage while shorter stockpiling periods produce higher quality forage but lower yield.

If we start from the idea that a square foot of ground can only capture so much solar energy, it becomes apparent the same square foot of ground can only support a finite quantity of forage growth. In typical late summer-early autumn growing conditions, it takes fescue about 60 to 75 days to reach its peak yield potential. Some other species such as orchardgrass or timothy will reach their peak yield in fewer days. If we stock-pile a pasture for a longer period, we end up with approxi-mately the same amount of yield, but lower quality forage.

A classic piece of Missouri stockpile research from the late 1960s, looked at stockpile yield from fields rested begin-ning in mid-June, mid-July, mid-August, and mid-September. The June, July, and August stockpiles all had the same yield while the September stockpile was significantly lower yielding than the first three. The June and July stockpiles had signifi-cantly lower crude protein and digestible energy compared to August. September had the highest nutritive value of any of the stockpile.

For dry, pregnant spring-calving cows, our target for stockpiled forage is usually maximum yield. We can expect dry-cow quality feed even at maximum yield. We need 60-75 days of growth to reach maximum yield. The way you determine when to begin stockpiling is to estimate your last day of grass growth and back up 60 to 75 days. In north Missouri, we considered November 1 to be the end of our grass growing season so we would begin stockpiling around August 15. Where we currently live in Idaho, grass growth is pretty well finished by October 1, so the target starting date is around July 15.

In a grazing situation, not every paddock is going to start stockpiling from the optimum date. Some paddocks you will graze for the last time 80 to 90 days before the end of growth while others may only have 50 days of rest. That's just the nature of rotational grazing. If you are accumulating stockpile on hay fields, it is much easier to have most of your acres starting closer to the optimum date.

Every year is going to be different due to moisture and temperature but this process gets you in the window of optimum stockpile growth. We have measured tall fescue still actively growing in early December, but the daily growth rate is quite slow compared to what it had been in September and October.

If the stockpile is planned for growing young stock or lactating females, shortening the stockpiling period can significantly increase nutritive value of the available forage. The increase in forage quality comes at the price of reduced forage yield. We have run fall calving cows the entire winter on nothing but stockpiled pasture and had excellent performance. The key factors with lactating females on stockpiled pasture is to make sure they start the winter in good body condition and don't force them to eat every last scrap in the pasture.

Young stock are more sensitive to forage quality than even lactating females, so their condition and performance needs to be closely monitored throughout the winter. Young stock are much more likely to require some form of supplementation to perform acceptably on stockpiled pasture.

How to begin the stockpiling process:

You want to start your stockpile growing from the best possible conditions to produce a high yielding, high quality stockpile. You want to grow forage, not weeds. You want to have mostly vegetative regrowth in the stockpile, not accumulated dead stems.

To accomplish this, start by grazing or clipping the pasture to about a three inch residual prior to target initiation date. If it is a pretty clean pasture and has been appropriately grazed for most of the season, you can start right from the last grazing residual. Your goal is to have an efficient solar panel going into the stockpile period. If a pasture has been grubbed down short all summer, it won't stockpile very well because it will take too long for the initial recovery of leaf cover. The gate may have been closed on August 15 but the pasture may not have really started regrowing until September 15.

If the pasture is weedy or has a heavy accumulation of dead forage stems, mob stocking or clipping the pasture is a good idea. Target residual height should be three to four inches. Taking it shorter or leaving it taller will detract from having an efficient solar panel. If you leave the dead stuff standing, it will carry forward into winter and the average stockpile quality will be lowered.

For the pasture to grow vigorously in the late summer and fall, it needs to have available water and nitrogen. In natural rainfall areas, it is the amount of rain received in August and September that really determines the final stockpile yield. If you have received adequate precipitation in August and there is water in the soil profile, stockpile growth takes off much better. It's the basic rule of grass grows grass. The sooner the grass canopy closes and captures most of the incoming solar radiation, the more rapidly the pasture grows.

On irrigated land, late season water availability is critical for growing high yielding, high quality stockpile. As a rule of thumb, every inch of water applied in August and September will produce 400 to 500 lbs dry matter forage per acre. If you can only put on four inches of water, you will end up with 3/4 to 1 ton of forage with either tall fescue or meadow bromegrass. If you can apply eight inches of water, you should produce 1.5 to 2 tons of stockpiled forage.

The need for nitrogen:

I once heard the definition of an agronomist as someone who never ceases to be amazed that nitrogen makes grass grow.

Yes, nitrogen makes grass grow and to have a high yielding stockpile, there must be nitrogen available in the soil. That nitrogen can come from soil organic matter, legumes in the pasture mix, applied manure, or commercial fertilizer. Every ton of grass forage at 15% crude protein contains between 45 and 50 lbs of nitrogen. If you want to grow a two-ton stockpile, there has to be at least 100 lbs of N coming from somewhere.

Healthy, vigorous soils with organic matter greater than 3.5 - 4 % can supply a substantial amount of this N. Unfortunately many of our agricultural soils have less than 2% organic matter, particularly through the rolling country of the Upper South and Lower Midwest where soil erosion has taken its toll. Most irrigated Western pastures, particularly those on sandy soils, also have low organic matter. Where you are most likely to find an adequate supply of N coming just from the organic fraction in the soil are long-term grass farms with a history of excellent grazing management and legumes use or with long term applications of manure.

Soils that have been under a grass-legume pasture for many years and have been grazed appropriately typically show low response to N fertilizer application. If 30 to 50% of the annual pasture production comes from a legume component, N fertilization generally doesn't pay. Legumes differ in their capability for N fixation with alfalfa and white clover having the highest capacity, red clover and birdsfoot trefoil at intermediate levels, and lespedeza having the least amount of N-fixation.

Applying poultry, feedlot, hog, or dairy manure can also provide N for stockpile growth. Using manures for stockpile fertilization also has the advantage of providing a long rest period between application and time of grazing use. Sometimes livestock will reject pasture that has been recently treated with manures. That is rarely ever a problem on stockpile fields.

The disadvantage of using manures for stockpile fertilization is it is usually warm to hot weather when the manure needs to be applied. This can lead to much higher N losses as ammonia if it is raw manure or slurry. The odor is much worse and if you are near non-agricultural neighbors you can receive a lot of complaints and maybe get hit with a lawsuit. Think before you use manures. Using composted manures greatly reduces both the gaseous N loss and any potential odor problems.

Commercial N fertilizer applied in late summer for winter pasture production is often the best return you can make on a dollar invested in fertilizer. Spring applied N produces a lot of grass but it is at a time when you generally already have a lot of grass. The late season application gives you grass when you don't normally have it. A pound of standing grass in December has much greater economic value than a pound of standing grass in June because the December alternative is high priced hay while the June alternative is just another pound of cheap grass.

Ammonium forms of N are preferred to urea (46-0-0) for summer applications. Urea fertilizers can rapidly break down to ammonia and escape into the atmosphere if they are not rapidly incorporated into the soil. As most pasture fertilizer applications are surface broadcast, there is no incorporation except through dissolution by water and infiltration into the soil.

If you have to use urea, avoid making applications early in the day. An evening application ahead of heavy dew fall will generally get most of your N into the soil. Applying urea ahead of forecast rain is a good option as long as it is not a downpour that generates a lot of runoff. Urea applied ahead of sprinkler irrigation is usually a safe bet.

Ammonium nitrate (34-0-0), diammonium phosphate (DAP 18-46-0), and ammonium sulfate (AMS 22-0-0) are more stable forms of N that do not require immediate soil incorporation to avoid loss. If the pasture also needs phosphorus, DAP is a good choice. If you are in an area with known sulfur deficiencies, try AMS as part of your N source.

How much N to apply depends on soil type and moisture conditions, as well as how agressively you plan to manage the stockpile. Sandy soils need more applied N due to their low organic matter content, but unfortunately they have less capability to bind N so potential N leaching into the groundwater is greatly increased. I advise sacrificing some yield potential for the benefit of water quality safety. Applications for sandy soils should be in the 40 to 60 lbs range.

If you really believe you can grow more stockpile than the 40-60 lbs will produce, apply 40 lbs at the beginning of the stockpile phase and another 40 lbs a month into the stockpile growth period. Soils ranging from silt loam into the silty clays and clay loams can hold more N so the rate may be higher on these soils. Generally, I would recommend 60 to 80 lbs for these soils. On any soil type I recommend the lower rate in the recommendation range if the summer has been dry and use the upper rate if there has been good precipitation through the summer.

The more intensively the stockpile grazing is managed, the greater the harvest efficiency will be. You will harvest more forage per acre with daily strip grazing compared to three days compared to ten days and so on. The more of the produced forage you capture, the more you can afford to produce. In some research in Missouri, we saw near linear response of tall fescue up to 100 lbs-N/acre.

In 1990 I never would have recommended 100 lbs-N to any farmer or rancher for stockpile production. If I am confident the graziers will manage the strip grazing aggressively enough, I will recommend up to 100 lbs of N for stockpiling predominantly tall fescue stands with reliable precipitation or irrigation.

Whenever you evaluate the economics of N-fertilization for stockpile production, you always have to compare the cost of the stockpiled grass to the next lowest cost feed alternative. For most livestock producers, that next best alternative is hay. Here is a rule of thumb for estimating the relative value of N for stockpile vs buying hay. If the N efficiency is 20 lbs of forage for each pound of applied N, the price you can afford to pay for N fertilizer per pound is 1% of the value of the hay. For example, if hay costs $60/ton, you can afford to pay up to 60 cents/lb for N fertilizer.

From one source or another, stockpiling pasture has to have adequate N available to produce enough forage to make it worth grazing. The lowest cost stockpile generally comes from a mixed grass-legume pasture with 30-50% of the annual production coming from the legume component and at least 30% fescue in the mix for the East and 30% tall fescue or meadow brome in the West.

What about legumes for stockpiling?

Legumes differ in how well they stockpile and maintain feed quality. Alfalfa might fix the most N and have the highest standing yield at the end of the growing season, but it loses feed quality rapidly in the face of freezing weather. Frosted leaves quickly drop from the plant and soon all you have left are stems, and alfalfa stems tend not to be very digestible due to high levels of lignin. Does that mean never stockpile alfalfa? No, it just means use it on the front side of winter before all the leaves are gone.

Sainfoin has been tried in the West for stockpile but shares the same problems as alfalfa of rapid leaf loss and low stem digestibility. It can, however, make a significant N fixation contribution during the growing season and is a useful legume in both dryland and irrigated pasture mixes.

Annual lespedeza has the same problem as alfalfa in that it rapidly loses its leaves with the first frost, and sometimes before frost if it is an older cultivar susceptible to the many leaf diseases that affect lespedeza. Compound the low quality with low N fixation capability and lespedeza leaves a lot to be desired as a companion legume for stockpiling fescue. Its redeeming virtue is that it will thrive on a lot of soils where no other legume will even survive. It goes back to the old proverb, "The low quality forage you've got is still better than the high quality forage you don't have."

White clover makes very little fall growth and contributes little to stockpile yield except on wet soil sites. Another problem with white clover pastures is they tend to become overgrazed and spotty. If 40% of your pasture surface is short-grazed white clover, there won't be much other pasture out there to stockpile. As a minor component in the pasture, white clover is fine, just don't expect it to be a major contributor to winter grazing.

Birdsfoot trefoil in pure stand rapidly loses its leaves and becomes nearly unpalatable in the fall. But in a mixture it seems to hold its leaves much better and its stems do not become near as wiry. We have stockpiled a lot of pastures with trefoil as a 15 to 25% component and had animals fully utilize it and perform very well.

In my Midwest experience, red clover consistently provides the best companion legume for stockpiling. It makes good fall growth, produces adequate amount of N to boost grass growth, it does not weather as badly as alfalfa and lespedeza, and the stems remain fairly digestible right through the winter. Because it can flower and set seed even in the fall of the year, it also has the capability of reseeding itself following a stockpiled winter.

While in Missouri, we found alsike clover to be a wimpy plant with little value except in wet years on certain soil sites. However, it is an entirely different plant in the West whether it be under irrigation or in natural subirrigated meadows. On our Idaho pastures, it is a very productive and persistent plant. We find it does very well as a component in stockpiled pastures. Leaf retention is good and stem digestibility remains high. Here in Idaho, alsike clover is my first choice for a companion legume for stockpiling.

Other legumes that have been used in various parts of the country in stockpile situations bear comments. Winter annual legumes such as crimson clover, arrowleaf clover, and hairy vetch can be grown in stockpile situations but they contribute little to forage supply on the front side of winter. They can be very productive in the spring after the grass stockpile has been grazed. Some will reseed themselves for a few years, but most need to be reseeded each fall to ensure having them the following spring.

Berseem and ball clover are a couple of lesser known annual clovers that can make significant yield on both the front and back side of winter. Ball clover is best suited for the upper and mid-South, while berseem clover has been used in the Midwest and the Pacific Northwest, as well as the Southern states. Both of these clovers have moderate leaf retention and relatively high stem digestibility in the winter.

Nutritive value of stockpiled cool-season pastures:
Many people are really surprised when they learn what high quality stockpiled pastures can actually be. Of course, the quality can also be pretty low. It all depends on how the stockpile was grown. When a full season's growth is allowed to accumulate, a high percentage of the standing forage is seed stems. There are also leaves from early in the season that have died and withered and much of their nutritive value is gone, particularly in wetter environments. Because of the shading from the old growth, very little new growth occurred later in the season. This kind of stockpile may have crude protein level less than 6% and digestibility under 50%. In other words, it wouldn't even let a dry, pregnant cow maintain her body condition. She would be going steadily downhill.

This type of forage still has nutritional value to the animal if properly supplemented. Frequently a couple of pounds of protein supplement will be enough to let the rumen microbes extract enough additional energy from the coarse fiber to meet the maintenance needs of the dry, pregnant female.

Managed stockpile as we have talked about growing over the preceding pages can be quite a different story. A properly grown N-fertilized tall fescue stockpile may start the winter with 16-18% crude protein and digestibility close to 70%. This is adequate nutrition for any class of beef animal or smaller ruminant. It would even keep a moderate producing dairy cow ticking along nicely. Feed quality will begin to deteriorate as winter progresses and by the end of winter most stockpiled pasture is down to cow maintenance quality. Stockpiled tall fescue may still be lactation quality feed even at the end of winter.

We have monitored the change in forage quality of various stockpiled pastures throughout the entire winter. Typically we see fescue stockpile ending up at 9-10% protein and digestibility in the mid-50% range. Mixed grass legume pastures may start the winter with even higher forage quality than N-fertilized fescue, but the mixed pastures deteriorate at a more rapid rate than fescue alone. By the end of the winter, the two pasture types will be of similar quality.

Utilizing stockpiled cool-season pastures:

Most stockpiled cool-season pastures can be utilized to 70-80% of the standing forage. We still need to maintain adequate plant residual to keep pastures healthy and vigorous in the spring. Even in the winter, that should still amount to 500 to 1000 lbs dry matter/acre or about two to three inches. Tall fescue dominant pastures can be grazed more severely than most other pastures and remain vigorous. In general species with leaves close to the ground (fescues, bluegrass, ryegrass, orchardgrass) may be grazed shorter than those with elevated leaves (smooth brome, timothy, western wheatgrass).

Stockpiling cool-season grass-legume mixtures can provide some of the lowest cost winter pastures available. Keys to success are 1) choose the right pasture, 2) rest it for the appropriate length of time, 3) start from a clean stubble, 4) make sure adequate N is available, 5) and then graze it effectively.

Chapter 14
Stockpiling warm-season perennial pastures

Whereas cool-season grasses form the basis for most of the winter grazing in the regions north of Interstate 40, perennial warm-season grasses are the dominant forage south of I-40 and they are an important component of year-around grazing in the Southern states.

One thing to recognize right up front is warm-season forages tend to have much lower protein content and overall nutritive value in the winter than do cool-season forages. One of the positive aspects of warm-season grass pastures is cool-season annual forages can be interseeded into some of those fields for extending the grazing season on warm-season dominated acres.

Much of what was said in the previous chapter regarding stockpiling cool-season perennial forages also applies to stockpiling warm-season perennial grasses like bermudagrass, bahiagrass, and dallisgrass. Because of their lower inherent forage quality, there are some management adjustments that can be made to help alleviate these challenges.

There is a wealth of information available from Land Grant Universities all across the South for management recommendations for specific areas and local conditions. Almost all Southern states have conducted research and demonstration projects with stockpiled warm-season grasses. Texas A&M and Oklahoma State University have probably done the most work along with The Noble Foundation in Ardmore, OK. R.L. Dalrymple was a pioneer of grazing stockpiled bermudagrass long before it became fashionable. From Florida to Texas and north to Virginia and Kansas, farmers and ranchers have proven the economic value of winter grazing stockpiled bermudagrass and other summer perennials.

Choosing the right pasture:

Just like with cool-season grasses, some warm-season grasses work better for stockpiling than others. With bermudagrass and bahiagrass being the dominant grasses in much of the South, most attention has been given these two species. Bermudagrass is vastly preferred for stockpiling as it has higher quality and palatability relative to bahia. Some cultivars of bermudagrass are superior to others for stockpiling because of greater late season growth.

As a general rule, the more cold tolerant bermuda varieties, such as Tifton 85, are better choices for stockpiling. That is not to say that plain old Coastal won't work if that is all you have to work with.

Stockpiled bahia will rarely be more than dry cow maintenance feed and will almost always require a protein supplement. If the stockpile has a high percentage of dead, brown leaves, the cows may not even want to eat it. Tight strip grazing with a little protein supplement will improve their attitude. Winter annual legumes such as crimson, arrowleaf, ball, or berseem clover in the mixture can also help both nutrition and palatability.

Dallisgrass is a fairly common species in many mixed pastures but is not often grown as a monoculture like the two dominant grasses. Dallisgrass makes excellent late season growth and holds its quality reasonably well. Across the Upper South there are a lot of acres of mixed tall fescue and dallisgrass that make excellent stockpile pasture.

As with stockpiling cool-season pastures, avoid fields with heavy summer annual weed infestation. Summer growing broadleaf weeds contribute little to forage value and summer annual grasses can tie up a lot of N and reduce growth of the perennial pasture. On the other hand, some of the winter annual weeds such as mustards and other green rosettes can provide enough protein supplementation to help digestibility of the stockpiled warm-season grasses.

When to begin stockpiling:

Because warm-season grasses lose quality more rapidly with age than do cool-season forages, we generally don't want to let the stockpile grow as long with warm-season grasses. For bermudagrass, generally plan on about 45 to 60 days of re-growth. For the Upper South, this generally means you should begin stockpiling in late August to early September and for the Deep South early to mid-September. This will provide an optimum balance of forage yield and quality. Stockpiling for longer periods can lead to substantially lower forage quality and the need for greater levels of supplementation.

Bahiagrass quality is even more sensitive to the duration of the stockpiling period than is bermudagrass. Resting for only 40 to 50 days results in reduced yield but keeps nutritive value near dry cow maintenance requirement. Longer rest periods mean more supplementation will be required.

How to begin the stockpiling process:

For either bermudagrass or bahia, plan to have the pasture grazed or clipped to about two-inch residual at the beginning of the stockpile period. Grazing is the preferred method as clipping can leave a lot of thatch on the ground if there has been excessive growth. Too much dead thatch accu-mulation results in reduced growth rate and lower stockpile yield. For dallisgrass plan on leaving three to four inches of residual to begin the stockpile process.

Always remember the cleaner the field is when you begin the stockpile process, the higher the nutritive value of the forage will be. If you carry a lot of dead material through summer and don't clean the pasture off, you will have low quality forage when winter rolls around.

The need for nitrogen:

Just like cool-season grasses, nitrogen makes grass grow and all the same guidelines for timing and rates of N presented in the previous chapter apply here. Legumes, manure, and commercial fertilizer are all viable sources.

Not very many seeded warm-season grass fields in the South contain enough summer growing legumes to provide adequate N for stockpiling. A few Southern grass farmers who divorced themselves from the N-fertilizer paradigm a number of years ago have been able to raise the legume content high enough to eliminate the need for any N fertilizer for stockpiling. It took them several years to get to the 30-50% legume level.

The introduction of more heat tolerant white clover cultivars such as Durana and Patriot have provided more reliable summer clovers for the South. While legumes seem to grow almost by accident in other parts of the USA, paying attention to soil pH, phosphorus, potassium, sulfur, and boron levels is critical to maintaining legumes on highly weathered soils common throughout the South.

Soil testing is an essential part of legume management in the South. If all the other soil nutrients are taken care of, N fixation rate will be significantly improved and the need for commercial fertilizer can be eliminated.

Winter annual legumes oversown in the warm-season grass fields when the stockpiling begins can make a significant contribution to both stockpile yield and quality. Most of the winter annual legumes provide the majority of their yield in the spring. Ball and berseem clover are two that can provide yield enhancement in the fall and early winter as well. Hairy and common vetch can behave as either summer or winter annuals. Because they can grow through the summer in some soil and landscape situations, they can provide rhizobia-fixed N from the outset of the stockpiling period.

The widespread presence of poultry houses in many of the warm-season grass dominant areas provides the opportunity for poultry litter as an excellent N source for stockpiling. Either raw or composted manure can be used effectively. The odor associated with raw poultry manure is the greatest deterrent to its use in late summer if you are close to a town or a subdivision. Check local ordinances and general neighborhood relations before applying raw manure in late summer.

Applied N as manure or fertilizer will increase both crude protein and digestibility of the stockpiled forage. For bermudagrass or dallisgrass plan to apply 60 to 80 units of N about 45 to 50 days before the end of the growing season. Bahia takes a little less N so plan for 40 to 60 lbs of N.

How to utilize stockpiled warm-season grass pastures:

Stockpiled warm-season perennial pastures are primarily dry cow feed unless there is a strong legume or winter annual weed presence in the pasture. If you run a winter or spring calving or lambing operation, stockpiled warm-season grasses perform well as post-weaning to late gestation forage. With a fall-calving or stocker operation, overseeded winter annuals are a much better grazing option.

Stockpiled non-native, warm-season grasses do not have the ability to tolerate weather damage to the same extent as many of the cool-season grasses do. These pastures should be utilized by mid-January in most of the region where they can be grown. The more autumn precipitation typically received, the earlier in winter stockpiled warm-season grasses should be utilized as wet winter weather is the enemy of stockpiled forage.

Because of the limited protein content of frosted warm-season grasses, tight strip grazing is highly beneficial for maintaining protein-energy balance. If cattle have full access to entire fields of stockpiled warm-season grasses, they will quickly glean most of the highest protein forage and leave you with protein deficient stem material early in the fall. Consequently, your supplementation costs will be much higher. Strip grazing is a low-cost means of doling out the good stuff along with the garbage.

Chapter 15
Winter grazing on the tall grass prairie

When we think about those vast grasslands stretching across the central USA and Canada that greeted the first explorers, we also have visions of massive herds of bison, elk, antelope, and deer stretching as far as the eye could see. Sometimes we forget there were prairie glades all the way back through the Appalachian Mountains to the Atlantic coast. Native grasslands consisting of mixed cool-season and warm-season grasses and a multitude of forbs provided year-around grazing for these native ruminants from Maryland to Montana and Manitoba to Mississippi.

Beginning about the 100th meridian West, the grasslands began to change from tall grass to mid-grass and finally to the short grass steppes of the Rocky Mountain region. This chapter will look at the tall grass prairies of the eastern half of the USA and the next chapter will move West into the native range country beyond the 100th meridian where annual precipitation levels decline precipitously.

The forage supply and quality of the prairies were obviously adequate for the native ruminants to thrive, otherwise they would not have been there in such great numbers. One of the most important things to remember when we consider the presence of vast herds of wild ruminants on native grassland is their breeding and birthing cycles were attuned to the grass growing season. Offspring were born after spring growth had already begun and females were bred later in the season after they had regained body condition lost over the winter. The nutritive value of dormant standing native forage was adequate to maintain them through the winter only because they were at their lowest nutrient requirements. As we plan for year-around grazing on native grasslands, it is critical we remember how our livestock interface with the grass cycle.

In the eastern portion of the tall grass prairie region, abundant rainfall usually results in vigorous summer growth. As long as not all the pastures are grazed down during the growing season, some pastures can be carried forward into the dormant season as stockpiled pasture. Managing for a variable stocking rate through the year is just as important on the prairie as it is for managing tame grass pastures.

Where there is a good mix of both warm-season and cool-season species, stockpiled forage quality can be adequate for even late-lactation cows. Where there is a predominance of warm-season species, forage quality is more appropriate for dry cow maintenance. Remember plant diversity is the friend of year-around grazing.

Much of the prairie acres that still exist in the Midwest are invaded with non-native cool-season species, both grasses and legumes. While this may be offensive to many native prairie purists, it actually produces a more versatile pasture than pure warm-season grass prairie. On our farm in Missouri, we had several areas of remnant native prairie with non-native invaders. We had a good diversity of big bluestem, indiangrass, little bluestem, some paspalum species, and a little bit of switchgrass and greasy grass. Along with it we had timothy, red top, bluegrass, red and white clover, annual lespedeza, and about 10% fescue.

When we grazed these paddocks in the winter, we found that as long as there was some green cool-season forage in the mix, the native grasses were grazed uniformly and the livestock seemed satisfied. In the years when there was little cool-season growth, only the most desirable of the dormant warm-season grasses were grazed and the remainder was left largely unutilized and the cattle quickly became discontented. The green cool-season forage in the mix provided the protein supplement that allowed the rumen microbes to digest the warm-season grass fiber to gain the energy held within.

A prairie with a mix of warm-season and cool-season species can provide adequate maintenance, late-gestation nutrition for spring-calving cows. Remember spring officially starts March 22, not February 1. Any calving prior to the onset of spring growth is really winter calving.

The value of strip grazing for maintaining rumen balance on stockpiled forage is a recurring theme in this book. Livestock given free access to the entire pasture described above would have quickly selected all the green forage and left the coarsest stem material. With the green forage gone, additional protein supplementation would be required for the stock to utilize the remaining coarse forage. Routine supplementation is an ongoing cost only if your management system allows it to be. Moving electric fence is a much lower cost alternative.

Native prairie dominated by warm-season grasses can provide dry cow maintenance feed for the entire winter if appropriately stockpiled and grazed. Some situations may require protein supplementation as winter progresses. If cows are calving before the onset of grass growth, protein supplementation will almost always be needed.

One of the nice things about protein supplements is it does not have to be fed every day. Supplying the animals with protein every second or third day produces comparable performance to feeding them every day. There will be more discussion of winter supplementation in Chapter 25.

One of the big differences in grazing management between the native grasses and the non-native species discussed in the preceding chapters is the need for maintaining a taller post-grazing residual for the native species. More CHOs are stored in the stem bases of the tall grass species than most other grasses. Grazing them too short, even in winter, will reduce spring vigor and early growth potential.

The tall grass prairie region extends from the Southeastern states to the Sandhills of Nebraska. Across this expansive area the annual precipitation ranges from over 50 inches in the Southeast to less than 20 inches in the western Sandhills. With this wide range in precipitation both the potential winter carrying capacity and the opportunity for summer utilization vary tremendously.

In the driest parts of the prairie region, stockpiling a full season's growth for winter use is a common strategy. This allows the grasses to maintain a healthy root system and go into winter with a full bank of stored CHOs. Winter grazing is a healthy and sustainable way to use native grasslands in the drier regions.

Farther east, abundant rainfall allows both summer and winter use of the prairie. The native tall grasses require longer rest periods during the growing season than do the introduced cool-season grasses discussed in Chapter 13. Whereas tall fescue can be grazed several times during the growing season and still produce a high yielding stockpile, the tall grass prairie should only be grazed once or twice before stockpiling begins, depending on precipitation.

If you are in an area receiving 20 to 30 inches of annual precipitation, you might plan for one late spring or early summer grazing event and then begin stockpiling native grasses by mid-July. With a 30- to 40-inch precipitation regime, you may choose to graze once or twice in the growing season, but plan to begin stockpiling by early August.

If you receive greater than 40 inches of annual precipitation, you can most likely graze the prairie at least twice during the growing season and plan to begin stockpiling by mid August.

As mentioned earlier, the tall grasses require a taller stubble height than do the introduced grasses. This is true both during the growing season and the dormant season. If your prairie is dominated by big or sand bluestem and indiangrass, plan for a minimum of six to eight inch post-grazing residual. A little more residual won't hurt a thing.

If you have less dominance by the taller species and more little bluestem and sideoats grama, you will be tempted to graze shorter. Grazing shorter will further stress the limited amount of tall grasses and encourage the lower-stature species. If that is what you want, then graze to the three to five inch level. If you want to encourage the return of the taller species, you will need to leave a taller stubble.

I prefer to not winter on the same ground in successive years. On our Missouri farm we rotated our winter stockpile ground on a three-year cycle. Basically, one third of the farm was stockpiled each year for winter use. We did not have large tracts of prairie, but we had small remnant patches in each of those thirds and the native grass component was still increasing each year up until the time we sold the farm and moved West.

My main concern with wintering on the same ground in repeated years was the potential for soil damage. Because we were in an environment where the ground did not freeze solidly for extended periods of time, there was always a potential for mud anytime in the winter. Winter grazing on soft prairies can result in loss of stand density and vigor of the tall grasses. One winter did not seem to affect them very badly, but the few times in the early years where we did winter on the same ground in successive years, the negative effect was apparent.

Most of our farm was laid out in permanent paddocks that would allow about five to seven days of grazing when they were stockpiled. Within each of these permanent paddocks, we would strip graze the forage with one- to three-day breaks, depending on what else was going on at any particular time. Thus, we were never on a paddock longer than a week. If it became too muddy to allow cattle to go back across the already grazed strip we would move to a new paddock and return to the remaining forage when ground conditions permitted. Chapters 22 and 23 will look at stock water and fence strategies for winter grazing in more detail.

This is all part of the long term planning you need to consider as you design your winter forage systems. As with summer grazing, appropriately balancing use and rest of the land is an important part of maintaining health and vigor.

Chapter 16
Winter grazing native range

This is probably the most difficult type of forage resource for which to make broad guidelines regarding winter grazing. While we generally think of range as something occurring in the American West, we have native range in states all the way from Florida to Hawaii and north to Alaska. The basic definition of rangeland is non-arable grazing land where the forage is provided predominantly by native plant species. To add to the confusion, we have "improved range" which is non-arable grazing land seeded to non-native species. Whether or not the non-native seeded species are an improvement over the native species is still subject to debate.

Because native range occurs across such a diverse geographical area, let's start by looking at some of the challenges occurring in different rangeland settings. Variations in precipitation and temperatures create huge differences in winter grazing strategies on native rangeland. Productivity is obviously closely tied to annual precipitation. In the Southeastern states there is native rangeland that receives in excess of 50 inches of precipitation annually, while in the desert Great Basin, native rangeland may receive less than five inches annually. Obviously, these environmental differences call for different utilization management.

Winter grazing in Florida seems like it should be an easy no-brainer situation. Because the native range of Florida is predominantly warm-season grasses, the inherent challenge of low forage quality in the dormant season is a challenge. Fortunately, winter annuals grow readily throughout the mild winters in the Sunshine State. A small field of winter annuals grown adjacent to native rangeland can be used as a protein supplement source for the dormant range. Just a few hours of grazing on the annual pasture every other day provides enough protein to allow the livestock to extract adequate levels of energy from the warm-season grasses.

Using a combination of rangeland and supplemental winter annual pastures is a sound strategy for native grasslands throughout the Deep South. An ongoing challenge in the coastal areas is ground that is simply too wet for cattle to stand on. This is particularly true for tilled winter annual fields being used as grazing supplement. No-till seeding will help overcome this problem, but not in every situation. The shorter the time period livestock are allowed to use the annual pasture area, the less likely soil damage will occur.

Interseeding winter annual legumes directly into rangeland is a strategy to consider if maintaining the integrity of native warm-season range is not a concern. Given that a large percentage of Southeastern "native" range is already overrun with bahia and common bermudagrass, adding non-native legumes may be a little easier to accept. Arrowleaf, crimson, and ball clovers, as well as annual vetches, can work as interseeded annual legumes.

Moving west from Dixie, precipitation levels begin to decline rapidly. I've often heard it said that Dallas is the last Eastern city and Fort Worth is the first Western city. There is quite a bit of truth to that claim. Annual precipitation at Dallas is about 37-38 inches while just 35 miles away at Fort Worth it is only about 34 inches. Moving another 150 miles west to Abilene, annual precipitation drops to just 24 inches. That is one inch less precipitation for every 15 miles you travel westward. Running across the range a little farther north, precipitation drops at a rate of one inch per 20 miles from Topeka,

Kansas to Colorado Springs, Colorado. That is a tremendous change and has a huge impact on both the type of vegetation and relative productivity.

On the southern rangelands moving west from Fort Worth, snow is rarely an impediment to grazing. That is not to say the occasional blizzard or ice storm won't shut you down from time to time, but year in and year out, snow is not the problem. What really limits winter grazing on the southern rangeland is overuse during the growing season thus leaving nothing behind for winter use. The most common solution to this problem is feeding hay harvested from cropland fields at great expense. As in every other part of the country, there are savvy ranchers who understand the power of grazing and have figured out how to make a year-around grazing system work on rangeland. One of the keys is stocking the ranch to its winter grazing capacity and utilizing stockers to balance out forage supply when necessary. Calving in sync with nature is the other key.

Moving a little farther west into New Mexico, Arizona, and the southern portions of Colorado and Utah bring higher elevation. With higher elevation come shorter growing seasons and increasing likelihood of snow cover. Moving north up along and through the Rocky Mountains brings even shorter growing seasons and higher elevations. Snow cover deep enough to end all grazing opportunities is a reality as we move north through the mountains and the plains into Canada.

Deciding when to use what resources and landscapes is a critical part of planning for year-around grazing in the Northern Rockies and on the Northern Plains. Throughout these regions we can find ranchers who kicked the hay habit years ago and are well versed in the art of grazing in this challenging environment. Once again the keys are stock to your winter grazing capacity and calve in sync with nature.

Overgrazing during the latter part of the 19th century and through the first half of the 20th century resulted in the loss of many of the native plants of the healthy rangeland. Soil erosion from both wind and water erosion further depleted the topsoil and health of the range. In many areas great improvement has been made since the formation of the Soil Conservation Service in the 1930s and other range management projects. Unfortunately, most of our rangeland is not as productive or as species rich as it was 200 years ago.

Plant diversity is the friend of the grazier and his or her livestock. Each plant has a little different composition from its neighbor. Those subtle differences are what provided balanced nutrition for the vast herds of native herbivores that roamed the range before European settlement. Strictly from a standpoint of animal nutrition and well being, plant diversity is as important as the pounds of feed produced per acre.

This is where the difference between healthy native range and areas seeded with imported range grasses really stands out. Healthy native rangeland rich in a diversity of plant species may well meet the nutritional demands of dry, pregnant livestock while monocultures and simple mixtures of non-native grasses rarely do. Using native range for winter grazing and seeded range for spring and summer grazing may be a better fit for many ranches.

Many range scientists have come to believe the best way to improve the health and vigor of native rangeland in the semi-arid region is a full season of rest followed by winter grazing. However, another precept of range management is to vary the season of use on different range units. Rather than exclusively setting aside your best native range for winter use, you may want to alternate winter and summer use seasons or plan to winter one year in three on a particular area.

The problem with repeatedly using a particular range unit in the same season each year is that it allows the animals to select for the most preferred species at that time of year. Remember, one of the great advantages of a widely diverse plant community is there is usually something green and growing more days of the year than with simple plant communities. Once that preferred set of species disappears from the pasture, the very thing that made it a desirable pasture for that season has disappeared. By varying the season of use, selection pressure is taken off any particular set of species.

The problem with selective grazing is less severe with winter use because we are usually not retarding the regrowth potential of the plant as overgrazing does during the growing season. It is still possible to graze too short in the winter and reduce spring vigor of the abused plant. Strip grazing allows the grazing manager to better regulate residual heights of even winter dormant species and thereby minimize the risk of winter overgrazing.

Thus far we have talked about range with a sense of ownership where you can easily make a decision whether to utilize it in the season of your choice. The reality in much of the West is public grazing lands under either Bureau of Land Management (BLM) or the Forest Service (USFS) with a sprinkling of US Fish and Wildlife Service (USFWS), National Park Service (NPS) Bureau of Reclamation (USBR), Army Corps of Engineers (USACE), and several state agencies, and, of course, school sections.

For those who aren't familiar with the idea of "school sections," under the Homestead Act of 1862, settlers were permitted to stake a claim on 160 acres or a quarter-section of land. Each township consists of 36 sections so there could potentially be four families on each section of land or a total of about 120 to 140 families per township. Within each township, one section was set aside for support of the local school that was expected to be established as more settlers moved into that township. In many cases those sections were sold and the money used to build the school house and pay the school "marm." In other cases they were retained and leased to area farmers and ranchers to provide an ongoing source of income for the school district. Throughout the West, many of these school sections still exist from the original survey. The ranch unit on which we live here in Idaho includes a school section lease of foothills rangeland.

In Lemhi County where we live, 94% of the land is publicly owned. Most of the ranches in our community utilize public lands to one extent or another in their grazing operations. Public ownership severely limits the flexibility in use and management. In many cases, fear of impending lawsuits prevents public land managers from making intelligent choices when it comes to resource management. It is easier to continue mediocre management than to go through the paperwork and possible challenges to making a change that might rock the boat. It is an unfortunate state of affairs.

However, the bright spot in public lands grazing is that in some regional or district offices, the public land managers are actually concerned with doing what is best for the land and not necessarily what is best for their careers. In these cases, ranchers may be able to gain flexibility in management and use patterns that are not only beneficial for the health of the range but also for their ranching operations. There are entire books written on the pros and cons of public land grazing and I will leave it at that.

As potential range productivity declines with the westward decrease in precipitation, the number of acres required to winter a cow increases. Besides just the number of acres increasing, the length of rest required to grow that stockpiled feed also increases. As length of rest period increases and the forage is exposed to more weathering, the forage quality usually trends downward. This scenario sets up, perhaps more than any other environment, the need to have your cows calving in sync with the natural forage growth and quality cycles.

The ease of using temporary fences for strip grazing also goes down as the required acres per cow increases. This doesn't mean the need for managed grazing on winter range is any less than what is needed with stockpiled pasture in the Midwest. It just means our focus changes and the tools we use are different.

Tight strip grazing on stockpiled Midwest pasture is geared more towards achieving a high utilization rate than it is for forage quality management. On dormant rangeland, grazing management should be more concerned with maintaining healthy rumen function than strictly forage utilization.

On stockpiled fescue pastures we try to achieve about 70-80% utilization of the standing forage, sometimes even higher. On rangeland we may still not want to achieve anything more than 50% grazing utilization and frequently our target may be less, due to soil protection, water cycle, and wildlife considerations.

Because forage quality, both protein and digestibility, of stockpiled fescue is high, we don't worry too much about protein:energy balance. On the other hand, protein is often deficient or marginal on stockpiled rangeland. If our livestock can't get what they need out of the forage, then we are faced with the need to supplement the forage and that can become an expensive proposition.

All livestock are selective grazers when given the opportunity and nowhere is this more true than on dormant winter rangeland. If they have access to large areas, they are very good at gleaning the highest quality material in a relatively short period of time. Even though a rancher might have enough acres with enough standing forage to take a cow herd all the way through winter, rarely do they accomplish a full winter of grazing using set stocking.

The main reason for this failure is selective grazing, which usually leaves a large mass of even lower quality forage than what was there to begin with. The simple way to overcome this shortcoming of rangeland is to allocate it out in increments that correspond with the rate of passage of forage through the rumen and the frequency with which the rumen microbes require a protein boost.

Moving livestock every day on rangeland would be the ideal, but the scale of operations required to accomplish this make most ranchers back away. Moving every three days matches up well with rate of passage and provides an acceptable frequency of protein supplementation. For traditional "just turn 'em out" ranchers even this may be intimidating, but such a grazing strategy is readily doable with a combination of permanent and portable fencing. We'll talk about the details of layout and tools to use in Chapters 20, 21, and 22, but here's a preview.

If you have a herd of 300 dry, pregnant beef cows and your stockpile range has 20 CDA of grazable forage available, it should take about 15 acres to feed the herd for one day. If you have six sections (640 acres/section X 6 sections = 3840 acres) set aside for winter grazing, that looks like it should provide grazing for 160 days. In actuality it probably won't, due to weathering losses and wildlife use. If we allow 4% weathering loss per month for five months, we should have about 130 days of grazing available. Wildlife use is highly variable from ranch-to-ranch and region-to-region so it is nearly impossible to make a meaningful prediction of what someone might experience. In this example, we'll take off another 15% for wildlife consumption, leaving us with about 110 days of expected grazing for our 300 cows.

Using three days as a reasonable grazing allocation for rangeland, we find we need somewhere between 35 and 45 increments. The three-day forage allotment would be about 45-50 acres. If each section is fenced with barb wire fence into the usual one square-mile pasture units, your first thought might be running six or seven one-mile polywire strips across each section. I'm pretty sure you would get tired of that after the first couple runs.

Using a combination of permanent and portable fence makes the job much easier. Splitting each section down the center with a permanent one or two-wire hi-tensile electric fence cuts the distance of portable fence to 1/2 mile, which is a much more manageable distance. Then with just two break wires, each half section could be grazed in three strips of a little more than 100 acres. Where the forage is a little better, make the allocation a little smaller and make four or five strips per half-section. It is exactly the same management we use in stockpiled fescue pasture carrying over a hundred CDA, just done on a different scale.

We have had several of our consulting clients who used this basic strategy to intensify management of their large range units. In some cases, they found the half-section paddocks worked so well, they split them lengthwise again to create four 160-acre paddocks each 1/4 mile wide and a mile long. Strip grazing these pastures required only 1/4-mile reels of polywire and fewer step-in posts so the daily chore time became even less.

One rancher commented that three years of this management made more range improvement than 20 years of traditional three-pasture rest-rotation they had been following religiously. Was he at first skeptical whether there was even a place for polywire in a range operation? You bet! The first thing he had to change was his mind. After that everything else was easy.

Stock water availability is often a limitation to winter grazing on rangeland. Solutions vary from running pipeline water across the rangeland to let them use snow. Every case is a little different and there is no one-size-fits-all solution. Chapter 22 will look in detail at providing winter stock water.

Rangeland across the country is highly variable in every aspect. You'll need to develop your own unique strategies to capitalize on your opportunities and overcome your challenges. Just remember, the first thing you have to change is your mind.

Chapter 17
Growing winter annual pastures

As the name implies, winter annuals are forages that are seeded in summer or fall, overwinter, and then produce their seed crop in the spring. From a forage quality standpoint, winter annuals are the highest quality feed we can offer to our livestock in late fall and early winter. In the South, where winter annuals sometimes continue growth throughout the winter, they can provide lactation and finishing quality forage throughout the winter.

Winter annual forages include many species of grasses, legumes, and forbs. Not all are adapted everywhere throughout the USA, but there is something from this group of plants that can be used at almost every location. While many of the annuals are mechanically reseeded each season, there are a number of species that are good natural reseeders.

In northern areas, winter annuals typically get established in the fall, but may make relatively little growth in autumn. Most of their production comes in the spring and usually begins a few weeks earlier than perennial pastures. In some locations, including high elevation irrigated regions, these plants can behave like summer annuals. They can be planted in spring and grazed in summer, autumn, or winter. The brassicas make the greatest late summer and fall growth and are often considered a special group of plants. This group includes turnips, kale, rape, canola, and swedes, among others.

This chapter will focus on the winter annual grasses and legumes while brassicas are covered separately in the next chapter.

Through the more central part of the USA, production in autumn is highly variable depending on the soil moisture and temperatures. As irrigated pasture on the Plains, productivity is high and very predictable. In the high natural rainfall environment of Missouri, we were able to plant a wide range of winter annuals. If we had favorable fall precipitation, forage yield before winter ranged from one to three tons/acre depending on species. In drier seasons, some did little more than get established even though we had a long fall growing season.

In the Southern states, using several different planting dates and plant species can provide high quality grazing throughout the winter. Available soil moisture is the primary determining factor for duration of winter grazing in the South although periodic frosts and freezes can shut down growth and limit yield potential. However, in most years winter annual pastures provide a very high quality alternative or supplement to stockpiled warm-season grasses.

Choosing the right species and varieties:

Small grains and annual ryegrass are the main players in the winter annual grass category. The small grains have different characteristics making each more or less attractive in different regions and for different applications. The main points to consider when selecting which of the small grains to use are site adaptation, winter hardiness, overall productivity, seasonal productivity, and relative forage quality.

For example, while barley provides both high yield and quality forage in northern regions, it is not very well adapted to areas south of I-70. Oats winterkill in the north, but may grow all winter in the South.

Where annual ryegrass can be grown, it is often the easiest to establish, most productive, and highest quality annual forage. Prior to the 1990s, we really only thought of annual ryegrass as a southern pasture option. Plant selection for winter survival has pushed the northern edge of annual ryegrass use into Canada. Even though it is still used as a cool-season summer annual in the far north, it wouldn't even be in use there had it not crept gradually northward with use as a winter annual. By the start of the 21st century in Missouri, we were using annual ryegrass on a regular basis and were finding an increasing number of varieties that would survive our winters.

The following table summarizes important characteristics of winter annual grasses. All of these designations are somewhat arbitrary and are site and management dependent.

Characteristic	Barley	Oats	Rye	Triticale	Wheat	Annual Ryegrass
Regional adaptation	North	Anywhere	Anywhere	Anywhere	Anywhere	South /Cent
Winter hardiness	High	Low	High	Moderate	Moderate	Variable
Yield potential	Moderate	Moderate	High	High	Moderate	High/Var
Forage quality	High	High	Moderate	Variable	Moderate	High
Fertility requirements	Moderate	Moderate	Low	Moderate	Moderate	Mod/High
Regrowth potential	Moderate	Moderate	Moderate	High	Low	High

Part of the variability in characteristics is cultivar dependent. There are different varieties for each of the small grain crops that have been selected for stronger grain or forage traits. Usually forage varieties have better regrowth potential following grazing and greater leaf-to-stem ratio resulting in higher quality forage. This in not to say the grain varieties cannot make good forage, they just have some limitations. We often used just bin-run feed oats as low-cost seed for fall pasture.

A number of winter annual legumes can either be planted with the small grains and ryegrass or overseeded by themselves into warm-season summer pastures to enhance the quality of the stockpiled forage. For the most part, winter annual clovers are best adapted in areas south of I-70 and, for some species, only south of I-40. Common, hairy, and purple vetch all have more winter hardiness and are used well into the Northern Plains. Vetches seem to grow in almost every environment and soil condition. On some of our eroded clay knobs in Missouri, lespedeza and hairy vetch were the only legumes that would grow there. Besides providing a boost in forage quality through their high protein content, most winter annual legumes are good nitrogen fixers and can provide substantial residual N for warm-season grasses the following summer.

Here is a table highlighting the characteristics of the most commonly used winter annual legumes.

Characteristic	Arrowleaf Clover	Crimson Clover	Ball Clover	Berseem Clover	Vetches
Regional Adaptation	Deep South	South	South	South/ Central	Anywhere
Winter Hardiness	Low	Moderate	Moderate	Low	Moderate
Yield Potential	High	Moderate	Moderate	High	Moderate
Forage Quality	High	High	High	High	Moderate
Fertility Requirements	Moderate	Moderate	Moderate	Moderate	Low
Regrowth Potential	Moderate	Moderate	High	High	Low
Reseeding Potential	Moderate	Low	High	Low	High

As with the grasses, there are cultivar differences, particularly with arrowleaf and crimson clover. There are not really any different cultivars among the vetches.

One of the ongoing challenges with legumes in humid climates is leaf and stem diseases. This is where you will find the biggest difference among clover cultivars. Because these diseases can do anything from completely wiping out a legume crop to just lowering digestibility, it pays to select a variety with high disease resistance. For example, "Apache" arrowleaf clover is the most disease resistant variety and is heads and shoulders above any other available variety so it should be your first choice.

There is no reason to only plant one winter annual variety in a pasture. Mixtures can provide a longer growing and grazing season than monocultures.

Planting oats with any of the other winter annual grasses usually provides more fall grazing as the oats establish very quickly and can make a heavy yield before killing frosts occur. An oats-ryegrass mixture is very nice as the oats provide good fall grazing, but then usually winterkill allowing the annual ryegrass to flourish in the spring.

Mixtures of winter annual legumes, particularly when interseeded into dormant warm-season grass stands, provide longer grazing windows and higher forage quality. Mixtures also usually reduce the risk of stand failures by having different germination timing and moisture requirements. Microsite variations in soil conditions also make mixtures more reliable than using a single species.

Establishing the pasture:

All of the winter annual grasses and legumes can be established without conventional seedbed preparation. Many farmers and ranchers still do a complete tillage process to seed winter annual pastures resulting in higher cost and more potential mud problems. There is nothing more discouraging than having a nice stand of expensive annual pastures and the ground too soft to utilize it. Tillage also often brings buried weed seeds to the surface and allows them to germinate.

No-till seeding lowers overall costs, reduces mud and weed potential, and conserves soil moisture. In the right conditions, annual ryegrass and cereal rye as well as many of the legumes can be broadcast seeded quite effectively. The right conditions means existing vegetation has been controlled either through grazing or chemical suppression, the seed is brought into contact with the soil, and there is adequate soil moisture for germination and establishment.

Broadcasting seed ahead of high stock density grazing is one of the lowest cost methods of seeding these pastures. An electric seeder on an ATV is an easy way of seeding many of these crops, although the necessarily high seeding rate for cereal rye requires frequent refilling of the hopper. Smaller seeded legumes can be seeded across many acres between refilling stops. Annual ryegrass is typically seeded at 15 to 25 lbs/acre so the refill frequency is intermediate.

Cereal rye and ryegrass are often spread with a fertilizer mix as an efficient way of getting the larger volume of seed out. This can be done ahead of high stock density grazing or on tilled ground followed by rolling or harrowing.

A variation on broadcast seeding is overseeding a winter annual mix into standing row-crops such as corn or soybeans. This is an increasingly common practice in the Midwest and on the Plains. Seeding is done either by aircraft or highboy about the time leaves are dropping on soybeans and corn leaves are beginning to dry out. This practice gets double duty out of crop land, increases the grazing value of crop residues, and captures residual nitrogen remaining in the soil following crop harvest.

No-till drilling works with all of the grasses and legumes. It is more expensive than broadcast seeding, but has value from the standpoint of consistency of success and evenness of stand. Because the small grains can germinate from much deeper in the soil than either ryegrass or the legumes, it pays to use a drill with at least two separate seed boxes. A common practice is to run the small grain seed into the seed drop going into the coulter or shank openers while letting the ryegrass and legume seed dribble out on the surface behind the opener through a disconnected seed tube.

If you are seeding just ryegrass and/or legumes with a no-till drill, one of the keys to success is not putting the seed too deep. There are many more stand failures from putting seed too deep compared to putting it too shallow. Of course the other key to success with no-till seeding pastures is to do it the day before you get an inch of rain. Seriously, watching the weather reports and getting the seed into the ground ahead of a rain usually gives much better results than waiting for it to rain and then drilling.

In an irrigated environment, plan to run the sprinklers immediately after drilling. Make sure you are putting on enough water to get the seed germinated and established. One common failure is putting on just enough water to sprout the seed and then letting the field dry out too much before returning with a second shot of water.

This is particularly a problem with farmers and ranchers who are used to running wheel lines on a 14-day return for growing alfalfa hay. A newly seeded pasture needs a much quicker return time.

It's a good idea to use seeding rates substantially higher than those normally recommended for small grain crops. High seeding rates will usually result in a thicker, denser stand that accelerates the development of your pasture solar panel.

Even with annual crops and pastures always remember your primary business is capturing solar energy. The quicker you get you solar panel up and running, the more productive that pasture will be. I like to use 50% more seed than recommended. While this may initially sound like an exorbitant spending spree, if adding another bushel of barley to the mix at a cost of $8/acre gives you another half ton of forage that fall, it will still be some of the cheapest feed you ever buy.

Soil fertility needs

Most of the annual grasses are very responsive to nitrogen. The winter annual legumes will fix their own nitrogen through symbiosis with the Rhizobia bacteria so they don't need N fertilizer, but they may need higher levels of P, K, S, B, and Mo. To get the most out of your winter annual pasture investments, make sure you have soil tested and know what the crop needs.

Adequate N may be available from high organic matter in a healthy, living soil. Most graziers have this as a long term goal, but most of us aren't really there yet so we look to other sources of N. Commercial fertilizer has historically been the quick-fix solution to the N question. With the ever increasing cost of N fertilizer and growing concerns over environmental impact, we need to look for other solutions.

Legumes growing on the same land in the summer have fixed N and, hopefully, raised the soil-N reservoir to an adequate level. Remember we are thinking in terms of cost effectiveness, not maximum yield. Winter annual legumes seeded in the mixture may provide N to the companion grasses later in the spring, but they will not help with getting the crop established.

Manure brought in from other farms is a good way to provide N for annual pastures. It is also a means of addressing deficiencies in other nutrients you may need to supply. Much of the South where winter annual pastures have the greatest potential also has a high concentration of poultry production. Many poultry farms have already saturated their soils with some nutrients and are not permitted to spread any more litter on their own property. Situations like this are ones to capitalize on as they can often provide the manure at much lower cost.

Good grazing management will retain most of the nutrients on your farm and within the pasture where it is grazed. That makes use of external nutrient sources good long term investments. Only nitrogen, sulfur, and boron are mobile enough to need ongoing replenishment. In many areas the sulfur levels are still high enough from the days of bad air pollution and acid rain to be able to provide S for many years to come. Make sure to get sulfur levels tested when you take soil samples.

Utilization of winter annual pastures

Because winter annual forages provide much higher quality forage deeper into winter than stockpiled cool-season or warm-season pastures, they can be used for extending the lactation season on pasture-based dairies, support fall-calving beef cows throughout the winter, and extend the finishing season for beef, lamb, and other meat animals. Winter annual pastures can be expensive to establish and maintain, but they have very high value relative to cost, if used effectively.

For the enterprises described above, the winter annual pasture may be the sole forage in their diet. While these high cost/high value pastures may be too much for just a dry pregnant beef cow or ewe, used as a supplement to stockpiled warm-season grasses, they can be very cost effective. As discussed with range utilization in the Southeast, just a couple hours of winter annual grazing every couple days can greatly enhance the rumen's ability to utilize low-protein warm-season stockpile.

Chapter 18
Using brassicas and other forbs

Brassicas are in a chapter of their own because they are in a class of their own. This does not mean they shouldn't be grown in mixtures with winter annual grasses and legumes, it just means they can be used very effectively on their own. Brassicas provide high yielding, high quality options for autumn grazing.

Because they can germinate and develop very quickly, they can provide lots of high quality forage in a relatively short window of time. This rapid development makes them a good shoulder-season crop between the time warm-season pastures are declining in growth rate and quality and before typical winter annual grass-legume mixtures may be ready to graze. They can also allow you to delay grazing of stockpiled perennial pastures until deeper into winter. In variety evaluation trials at FSRC, we had stemless kale seeded around September 1st yielding close to three tons per acre by late October.

Forage turnips and kale can often be grazed in as little as 30 days after seeding. If the new crop is lightly grazed and then rested, it can regrow very quickly and be ready for another grazing in 30 days or less. With strip grazing we usually plan to begin grazing at a much less advanced growth stage than if the crop were going to be continuously grazed throughout the fall.

If you delay the beginning of grazing until all the crop has reached an advanced stage, the amount of forage lost to freezing damage will likely be much greater.

Most of the brassicas, with the exception of rape and canola, are fairly small seeds so it doesn't take a lot of seed per acre. Turnips and kale are often added to mixtures at a rate of only a couple of pounds per acre. Even for a pure stand 4-6 lb/acre is plenty. Seed cost tends to be fairly low, making these species an easy addition to almost any mixture.

The bulb-type species like turnips and swedes do best in prepared seed beds as loose soil allows the bulbs to grow more vigorously. Kale and forage turnips can be no-till seeded either as broadcast or drilled seed. As with the legumes, drilling too deep is the biggest cause of stand failure with the smaller seeded brassicas.

Any of the brassicas can be used either by themselves or in a winter annual grass-legume mixture overseeded into standing crops like corn, grain sorghum or soybeans. The high protein and digestible energy of a brassica make a nice complement to lower quality corn or sorghum stalks. A corn residue field that was expected to be little more than dry cow feed can be upgraded to growing or finishing pasture by adding brassicas or other winter annuals to the field.

All of the brassicas do best on moderate to high fertility soils, although some can produce moderate yields on more marginal soils. Swedes seem to do better on lower pH soils than the other brassicas. Another advantage to seeding them on crop ground is their ability to take up any residual nitrogen remaining in the soil in the relatively short autumn growing season.

One downside of most of these species is they are not very frost resistant and forage yield and quality can disappear quickly once cold weather sets in. In most environments with real winter, don't plan for more than a month or two of use with brassicas or you will likely have a lot of forage go to waste. In milder climates of the deeper South, they may continue to grow all winter and be utilized for several months.

In north Missouri we used a mixture of cereal rye and

turnips for sheep grazing a couple of winters. We found we needed to have them utilized by mid-December otherwise both tops and bulbs quickly went to mush. This is where we found a mixture with a small grain or ryegrass was much more durable than a pure stand of turnips. We had used a broadcast spreader to seed the mix and had strips where the turnips were much heavier than the rye. Where it was turnip-dominant, forage deterioration was much quicker than where there was about a 50:50 mixture of rye and turnips. The lower the turnip density, the longer they held up with cold weather.

There are more cold hardy varieties of kale and some of the other species can withstand much more freezing weather and still maintain their integrity. Some of the improved kale varieties can withstand temperatures down into the lower teens.

By themselves, brassicas can almost be too high in quality. Crude protein content over 20% and digestibility over 80% are not uncommon. Moisture content is almost always greater than 90%. This is why they work so well in mixtures with small grains or with crop residues. The rumen needs some higher fiber content feed to slow down rate of passage.

Many farmers and ranchers have heard stories of cattle or sheep choking on turnip bulbs and may shy away from using brassicas for fear of death loss. I have talked to very few graziers who have ever actually experienced any death loss from this cause. When we used turnips for sheep, we would occasionally see one get a chunk lodged in their throat for a brief period, but they either were able to swallow it or cough it up and give it another try.

As with many other lesser known animal afflictions, I think the choking problem is something most people heard about from a neighbor's friend's uncle who saw it on his neighbor's farm in the 1950s. In other words, the value of using brassicas as part of your year-around grazing strategy is far greater than the potential losses.

By the way, the country remedy for dealing with a choking animal is shove the bulb down the rest of the esophagus with a hoe or broom handle. I never had to try that because we never had a problem.

There are a couple of realistic concerns with grazing brassicas of which we should all be aware.

If the entire diet is brassica, there can be rare occurrence of goiters and other metabolic disorders. The goiter issue is usually addressed by keeping iodized salt available. The neurological disorder rarely ever occurs if at least 20-25% of the diet is coming from other plant materials, hence another reason to plant a mixture with a small grain component.

There have also been mixed reports on brassica imparting off flavors to meat and milk. The off flavor has been much more commonly reported with milk compared to meat. Once again the best solution is simply planting a mixture with grasses and legumes rather than a monoculture. Off flavor in meat has been more commonly observed with lambs rather than beef.

Chapter 19
Stockpiling summer annual grasses and legumes

Corn, crabgrass and cowpeas are some of the diverse members of the summer annual forages. It might seem a little odd to plan to grow warm-season crops in the summertime to graze in the winter months, but the high productivity and potential yields make them an attractive option in some parts of the USA and even in Canada.

The tall growth stature and stiff stalks of corn and sorghum are some of the features that make them valuable as a winter forage in deep snow country. These are some of the very few plants that can stand up through three feet of snow and still be grazable. They can also be swathed to produce a really huge swath that holds up well under heavy snow cover.

The popularity of corn as a grazing crop exploded in the last few years of the 20th century and here in the beginning of the 21st century. A combination of low grain prices and the potential for a very high pasture yield per acre are the two primary factors driving the interest in grazing corn. Extensive field work by University of Nebraska extension educators Terry Gompert and Bob Scriven have shown the cost effectiveness of grazing corn. They have also developed grazing strategies to optimize efficient harvest of corn crops with grazing livestock.

There are two divergent strategies for growing corn as a winter grazing crop. The real farmer mentality is to make all the inputs necessary for growing a high yielding grain crop and

then just deviating from the norm by harvesting it with livestock. This "conventional" approach may use herbicides,
insecticides and commercial fertilizer to the same extent as
grain farming. The other approach is the "hippie" corn crop
where weeds are allowed to grow as a companion crop, no
insecticides are used (who cares if the stalk falls over?), and
with minimal to no fertilizer inputs. The potential yield is
usually much lower with the hippie approach, but the input
costs are minimal.

The same basic strategies can also be applied to growing grain or forage sorghum for pasture. Using a companion
legume like soybeans, mung beans, or cowpeas with corn or
sorghum crops can provide some nitrogen for the mixtures.
Another way to provide non-fertilizer N to the summer annual
crop is using a rotation with vetch or other winter annual
legumes the previous winter. When the winter annual legume is
grazed off in the spring, nitrogen fixed by that crop will be
slowly released throughout the summer to supply the summer
annual crop.

The forage strategy described above is obviously better
suited to a grazier who is already in a mixed crop and livestock
operation. I would not encourage any grazier with good existing
stands of perennial pastures to plow them up or spray them out
just to grow summer annual forages for winter grazing.

The sorghum family includes sudangrass, johnsongrass,
and cane and grain sorghums. One thing all of these plants have
in common is the possibility of prussic acid poisoning. Prussic
acid is a precursor to cyanide in the plant and is converted to
cyanide when the plant is stressed. While prussic acid is most
commonly associated with drought stress, it can also occur as a
response to frost. The good news is the problem disappears in
just a few days after the first killing frosts. If you plan to delay
use of the sorghums until cold weather sets in, there shouldn't
be any problems.

One of the challenges of grazing corn and sorghum that
may stand several feet above the head of cattle or other live-

stock is running strip graze fences through the tall crop. Many graziers use an ATV or pickup truck to knock down a path through the tall crop. With high stock density grazing, the livestock will usually do a good job of cleaning up the forage that was knocked down in the process.

Another option is to deliberately leave some skipped rows in the field to allow easy erection of the portable fences. This takes a little forward planning to be able to make your best estimate of how many rows or feet of the crop will need to be allocated for each move. A three- to four-foot open strip is all it takes to make an easy passageway for your fence. In some situations, particularly using the no-herbicide approach, weed growth in these open strips may be as much of an obstacle as the tall crop would have been.

Crabgrass is another summer annual grass that can be an important component in pastures through the central and southern parts of the USA. Although it is much better as a summer grazing crop, crabgrass can also be stockpiled and grazed in the winter. It works much better if it is in a pasture mix rather than by itself. The farther south we go, the greater its value as a winter forage. Severe frosting in the northern extent of crabgrass range rapidly degrades the forage to little more than straw. In a mixture with cool-season grasses and legumes like tall fescue and red clover, it is readily grazed and digested in the rumen.

An unplanned opportunity for many small grain farmers are the summer annual weeds that come up in stubble fields. This is another nice shoulder-season pasture that can help bridge the gap between the end of summer pastures and the new availability of winter annual pastures or stockpile grazing. Many of the summer broadleaf weeds like lambsquarter and pig weeds are readily grazed along with grasses like barnyardgrass, fall panicum, and green or giant foxtail during the late summer or early fall months. Utilizing some of these weedy stubble fields may buy the extra time you need to stockpile a few more acres of your permanent pastures.

Grazing summer weed growth in the fall is another situation where tight strip grazing can easily increase the potential carrying capacity by at least 50%. Some of the weeds are a little less than desirable after they have matured, but every day moves can keep the good, the bad, and the ugly all going into the rumen at the same time resulting in much better utilization of the bad and the ugly.

There has been much more attention given to summer annual grasses for winter grazing compared to summer annual legumes for several reasons. First, there are not that many summer annual legumes used for grazing. Second, the legumes tend to lose their leaves very quickly in late summer, even well before frost. Third, stem digestibility is very low on average. Soybeans are a classic example of why summer annual legumes are not very good grazing crops.

So why even mention them at all? Nitrogen fixation for the companion grass crop is one good reason. All of the summer annual grasses respond very well to N fertilization. As a more sustainable practice, we need to be looking for other options. Cowpeas seem to be the best N-fixers among the summer annual legumes with typical fixation rate reported to be 120 to 150 lbs N fixed/acre annually with some measurements as high as 300 lbs/acre. So even if they contribute little to the actual winter forage supply, they have made a strong contribution to the companion grasses.

The dried leaves of the various bean and peas that have fallen to the ground retain moderate nutritive value and can be consumed by livestock off the ground. This is a situation where tightly controlled strip grazing is imperative to achieving a high utilization rate.

All of the summer annual legumes have greater potential use south of I-40 due to later frosts and hotter summer growing conditions. Cowpeas have been evaluated as far north as North Dakota and have shown some potential even in that region, and soybeans are grown all the way into Canada. In the more northern regions, certainly anywhere north of I-80, the cool-season legumes offer much better winter grazing potential. Remember to build your winter forage supply around proven plant performers in your region, but don't be afraid to try a little experimentation to keep pushing the envelope.

Pigeon peas in the Pahsimeroi? Probably not.

Chapter 20
Keeping winter grazing simple

Most farmers and ranchers consider feeding hay to their stock to be a fairly simple task. You go to the bale yard, pick up some hay, drive out to the stock, and feed them some hay. Simple, but very expensive. Like almost everything else in life, convenience comes at a price. It is up to you to determine what the true cost of each alternative strategy might be and then decide if you're willing and able to pay the price for the option you choose.

Winter grazing can be as simple and as convenient as feeding hay, but much less expensive, if you plan to make it so. Trying to graze stock through the winter without a coherent plan and the right strategies for your operating environment can turn it into an inconvenient nightmare. In this chapter we'll look at strategies for keeping winter grazing as simple as possible. The next several chapters will look at each of these considerations in greater detail.

Obviously, one of the first steps to being able to graze year-around is knowing how much forage you need to provide to your livestock at different times of the year. Setting the stocking rate for your centerpiece enterprise based on your winter grazing capacity rather than summer grazing capacity is one of the first steps towards efficient year-around grazing. If financial analysis shows it is not economically feasible to operate at that stocking rate, it is then necessary to either cost effectively increase winter forage supply to be able to carry the number of head necessary to operate profitably or reconsider what your primary business should or shouldn't be.

You want to go into winter knowing what you will be doing through each month or week of the dormant season. If the end-of-the-growing-season pasture inventory shows you do not have enough pasture to go through the winter with all the stock you had planned, start making adjustments now. Do not wait until the standing forage is gone before you figure it out. Always remember there is a minimum of two ways to look at the situation. You can increase supply or you can reduce demand. The sooner you make your adjustments, the smoother your path will be.

Knowing your beginning inventory will make feed budgeting easier throughout the winter.

In some parts of the country, available stock water is a limitation to which pastures can be used during the winter, while in other areas livestock rely entirely on snow to supply their water needs.

Remember dry, pregnant cows or ewes have a much lower daily water requirement than do lactating females. Sheep have lower water requirements than do cattle. They can go for longer time periods between drinking and their absolute requirement relative to metabolic weight is lower than cattle. Cattle, particularly, can travel farther distances to water in the winter than the summer because of both lower temperatures and reduced daily requirement. Water availability in the winter is relatively less critical than in the summer months.

If you have limited winter water resources and snow is not a reliable source, planning around the available water sources is critical to success. Some sources may freeze up much earlier in the season than other sources. You need to be aware of what the typical patterns are and plan to save the pastures with the most reliable water sources for the coldest part of winter.

Within a pasture with a single water source, base your strip grazing strategy on starting close to water and moving away from it with successive strips.

Optimize the availability of stock water in your grazing plans.

In larger operations, the winter strip grazing program may only consist of moving from one permanent pasture to another. One of our range clients rotates 800 to 1200 cows through a series of 160-acre pastures every five to seven days. They use no movable electric fences, but operate what would be considered a very intensive winter grazing operation for semi-arid rangeland. Smaller operations are more likely to use movable fences to control the daily feed allocation.

What constitutes small is a matter of perspective. Here in Idaho we strip graze 300-400 cows using PolyBraid on reels and step-in posts. The length of fence for each move is about 1000 ft. When we were in Missouri we only grazed 50-60 cows in the winter and the fence distances were usually in the 300-400 ft range. Same basic principles, just different lengths of fence. The time required for moving the fence is still much less than required to feed hay to the same number of cows in either case.

Because you are typically working out in the cold when using winter grazing, keeping the lengths of fence to be moved to something within your comfort zone is important. Also, using fence products that work reliably and easily will help keep chore time to a minimum.

Plan your fence system to meet your comfort level for working in cold conditions.

One of the newer innovations in winter grazing that emerged in the 1990s was swath grazing. Early work was pioneered in the Canadian prairie provinces and has since spread into the USA. As the name implies, swath grazing involves one step of the hay harvest process, but ends there.

Forage is left in a swath and then grazed later in the winter. Swath grazing only works in the relatively dry parts of the country and is ideally suited to irrigated fields in semi-arid regions.

Swathing has advantages over simply stockpiling pasture in certain situations, but it does come at the added cost of mechanical swathing. Whether it is annual pastures or alfalfa fields, the crop can be swathed to lock in a particular quality stage. This is especially useful for growing and finishing operations or dairies. In dry environments, yield and quality losses are minimal with swath grazing. Swath grazing is also very useful in deeper snow country as it concentrates the forage in a smaller area and is more accessible by the livestock.

Consider swath grazing if you need to lock in forage quality or face deep snow situations.

Depending on the forage you have available and your livestock needs, supplementation may become necessary at certain times. The key word here is supplementation. By definition a supplement is something provided in addition to a basic input, not a replacement of the basic input. Knowing what your livestock need and what your pasture can supply are the first steps to planning supplementation strategies.

Supplementation should be kept as simple as possible so it can be done efficiently and cost effectively. When evaluating the cost of supplementation, there is more to consider than just the price of the feed itself. Feed bunks, hauling, labor, and wastage all need to be weighed against the realized benefit to accurately assess cost:benefit ratio. Another important concept to remember is protein supplementation does not have to occur every day.

Use strategic supplementation to maintain animal performance and stretch forage supply.

You will face years where the best of plans will not work. Winter grazing is predicated on the assumption that you will be able to grow some forage during the growing season to graze during the dormant season. What if there is no significant rainfall during the growing season? What if a Federal judge cuts off your irrigation water? What if grasshoppers of Biblical proportions ravage your landscape?

Every farm or ranch needs to have a drought plan in place to help them deal with these situations. It is virtually impossible to feed your way through a drought. Significant destocking is usually the most profitable or least damaging way to deal with droughts. Destocking does not necessarily mean selling out. It means removing the livestock from your pastures and range. You may retain ownership while shipping them to some other location. It may mean just a partial reduction in stocking rate. There are always options. The key is to have a plan.

Have a drought strategy in place to help deal with the inevitable.

Depending on where you farm or ranch, you will also face periodic challenges of snow, ice, wildlife depredation, among others. Some animals learn to cope with these challenges more readily than other animals. Some managers learn to cope with these challenges more readily than do other managers. I have always emphasized the need for flexibility in grazing management. Nowhere is this more true than in winter grazing operations.

Ongoing observation, analysis, and adjustments will be needed.

Chapter 21
Budgeting winter forage

The last several chapters have described many different options for providing winter grazing. The next step in the process is planning how that forage will be utilized. In a classic stockpile situation where there is zero forage growth over the winter, all the winter forage is produced in the growing season and then budgeted out over the dormant period. If you're planning to make it through winter without running out of feed or reaching spring with half of what you started with still standing in the pasture, you need to have a plan.

In Chapter 5 we talked about assessing your winter forage opportunities and charting when different options can possibly provide grazing. Chapter 6 showed how to use an ongoing pasture inventory to monitor how much forage is available for grazing at any given time. Chapter 9 covered how to put the pieces into a logical sequence of use. This chapter goes through the grazier's arithmetic needed to plan feed budgeting. We'll look at feed budgeting from determining total seasonal demand all the way through daily allocation.

Total seasonal demand is the product of animal type and weight, daily intake requirement, target utilization rate, and expected number of days of dormant season grazing. Depending on the class or classes of livestock that will be grazing, you need to decide whether to use an animal based unit, such as cow-days or standard animal unit days, or work with estimated forage yield as lb/acre and lb/hd/day. While the basic process is similar between the two systems, the flow of calculations is quite different.

If there is only one class of livestock on the farm in winter, using an animal-based unit, such as cow-day, is very simple. If there are several classes of livestock to be overwintered, using the standard animal unit day (AUD) or animal unit month (AUM) is advisable. If you plan to do custom grazing or are concerned about actual feed quantities, use the forage yield system based on lb/acre and lb/hd/day. We will look at examples for each of these strategies.

Some of these systems require you know the weight of the livestock, while others do not. A livestock scale is obviously the most accurate means of knowing what livestock weigh. If you own or have access to a scale, weighing all of your stock several times a year can be a useful and educational practice. When I was at FSRC, most of the cattle on the station were weighed every 21 or 28 days. When you are weighing livestock this frequently, your eyes can become well calibrated to estimating weight of critters standing in the pasture. While many people consider themselves to be very good at estimating cattle weights, the sad truth is most are not.

At many of our grazing schools, participants are required to estimate the weight of a set of cattle in a pen. There are usually four or five people on the team and usually their estimates of weight for four head of cattle will vary among the team members by at least 200 pounds per head. Underestimating the weight of your cows by 200 lbs is easily a 15 to 20% underestimate in feed demand. This can translate to 20 to 30 days of unexpected hay feeding and the associated cost.

If you are planning to weigh your livestock, it is important that you do it in a consistent manner. You should always plan to weigh livestock at the same time of day because their weight will vary tremendously throughout the day depending on whether they have just completed grazing, been to the water tank, or have been resting and ruminating. If they stand in a pen longer than an hour before crossing the scale, the last ones weighed will appear to be considerably lighter than the first ones weighed. The less stress you place on the livestock while you are working with them, the more accurate your weights will be. I prefer to weigh animals early in the morning before they have grazed very much or been to the stock tank.

Some of these systems may require a forage yield estimate in pounds per acre. In Chapter 6 a simple method for estimating forage yield based on the pasture composition, height, and stand density was explained. As with most of the measuring processes, if you do it consistently and follow the guidelines, the estimates you get will be fairly accurate. There are other means of estimating forage yield including rising plate meters and electronic pasture probes available from several grazing supply vendors. These tools are only accurate if they have been properly calibrated for your pasture conditions. If they have not been calibrated, the estimates they give you are no better than an untrained eye.

All of these methods for estimating forage yield per acre are based on all the biomass above ground level. It is not the quantity of forage above some theoretical optimum grazing or clipping height. It is above ground biomass. We do not expect our livestock to consume 100% of the standing biomass. We expect them to leave some residual forage behind to ensure the survival of perennial pastures or possibly future regrowth on an annual pasture. We know they will foul some forage with manure and trample other areas. This is where the utilization target comes into play. This is the percent of the standing forage we expect the livestock to harvest.

With winter grazing, the most important factor determining the degree of utilization is the period of time the livestock have to live on that piece of the dinner plate. While some people's first thought might be the longer you leave them on a pasture the more of it they will utilize, it is actually the opposite. The more days animals occupy the same piece of pasture, the greater the percentage lost to fouling, trampling, and bedding. The tougher the grazing conditions in terms of snow, ice, or mud, the lower the utilization percentage. The shorter the grazing period, the higher the total utilization rate will be.

Based on research at FSRC, these are the expected utilization levels for stockpiled or winter annual pastures for different grazing periods. On native range we still target only 40-60% utilization with winter grazing. Higher levels of utilization require more extended rest periods.

Length of grazing period	Expected % utilization
1 day or less	80
3 - 4 days	70
6 - 8 days	60
10 - 14 days	50
> 20 days	40

When the calculation calls for an estimate of daily intake level as a percent of body weight of the animal, there are a number of available publications and resources to help you out. One thing to remember, though, is it is always better to overestimate expected intake than to underestimate it. This is part of the buffer you can build into your grazing system to help keep you from running out of feed sooner than expected.

On the following page there is a basic table of approximate intake targets for breeding females of different species. These values are a little higher than what you would find in many university fact sheets. Most of those fact sheets are based on the National Research Council (NRC) requirements and don't allow for margins of error you will want to maintain in real-world grazing operations.

There is a wide range shown for peak lactation. This is due to the wide variation in milking potential within any specie or breed. Here is where it is important to know the relative milking ability of your stock. Since there can easily be a 50% variance in intake due to milking potential, you can come up really short on feed if you underestimate the lactation require-

	Beef cow	Dairy cow	Ewe/ Nanny
Production stage	**% of body weight**		
Dry pregnant maintenance	2.0 - 2.2	2.2 - 2.4	1.2 - 1.6
Late gestation	2.2 - 2.4	2.6 - 2.8	2.0 - 2.4
Peak lactation	2.5 - 3.5	3.5 - 4.5	3.5 - 4.5
Late lactation	2.4 - 2.6	3.0 - 4.0	3.0 - 3.5

ment of your stock. If you will only be winter grazing dry, pregnant animals, the likelihood of completely missing the boat is much reduced. Fall calving cows and fall lambing ewes are where you are most likely to get in trouble if your stock are above average milkers.

Let's look at some example calculations.

Single class of livestock using animal-based units:

This is the simplest example of a winter grazing program. This farm has 100 dry, pregnant beef cows and they typically experience 120 days of winter feeding. Their total winter forage demand is: 100 cows X 120 days = 12000 cow-days. They have been diligently conducting a bi-weekly pasture inventory throughout the growing season and the final inventory at the end of the growing season showed the 80 acres they stockpiled was carrying an average yield of 90 cow-days/acre (CDA). Thus, their stockpile supply is: 80 acres X 90 CDA = 7200 cow-days.

This farm will not be able to graze the entire winter because demand exceeds supply: 12000 cow-day demand - 7200 cow-day supply = 4800 cow-day deficit.

In this system there was no conversion to standard animal units, no question of what the cows weigh, and no calculation involving pounds of forage on either the supply or demand side of the equation. This system is based on experience with your herd and your pastures, nothing more. The best way to gain that experience is using the pasture inventory process on an ongoing basis to calibrate your eye to the needs of your herd and the production of your pastures. Compare your estimates from the pasture inventory to what the livestock actually harvested when they went through that pasture. Remember, during the growing season the pasture likely grew some more from the time you made your inventory estimate until it was grazed. Make an adjustment for growth during the calibrating process.

Single class of livestock using standard animal units:
Let's look at the same scenario using the standard animal unit system. A standard animal unit is the equivalent of a 1,000 lbs cow with calf at side. Strictly speaking, an AUD is the amount of forage that will be consumed by that animal on a daily basis and is generally considered to be 26 lbs/day. A standard AUM is 780 lbs of dry forage. If the cows on the example farm had weighed 1200 lbs, they would each be considered 1.2 AU.

The demand for forage is going to be: 100 cows X 1.2 AU equivalent/cow X 120 days = 14,400 AUD. If the pasture inventory had been conducted based on standard AUD, the total supply at the end of the season would have been estimated at 8640 AUD. The feed deficit is 5760 AUD.

What have we accomplished by using the standard AUD system? Nothing other than being able to talk in technical terms with public agency employees or the sheep farmer down the road who wasn't using EDA (ewe-day/acre) as his unit of measure. If you are not dealing with public agencies and aren't interested in speaking across specie and animal class lines, there is no advantage to standard animal units. Just stick with something that makes sense in your operation.

Single class of livestock using pounds of forage:
Using pounds of forage for both the inventory and animal side of the equations requires quite a bit more calculation. Some people like to use this system because it makes

them more comfortable with the answers they get. It must have something to do with the decimal point.

To use the pounds of forage based system you will need to know the weight of the livestock (lbs/hd), expected intake level (% of bodyweight), target utilization rate (% of standing forage), and have an estimate of dry matter forage yield (lbs/acre).

Step 1: Calculate individual animal requirement:
1200 lb/cow X 2.2% of bodyweight/day (.022 lb forage/lbs of cow) = 26.4 lb/cow/day.

Step 2: Calculate total feed demand:
100 cows X 26.4 lbs/cow/day X 120 days = 316,800 lbs of forage for the winter period.

Step 3: Calculate total standing forage:
Average pasture height of this tall fescue-legume mixed pasture is 10 inches and it is average stand density. Using Table 6.1 on page 40, we estimate the standing forage yield to be 10 inches X 350 lbs/acre-inch = 3500 lbs/acre.

Step 4: Calculate target grazable forage per acre:
The grazier plans to move temporary fence every 3 days so we select a target utilization rate of 70% 3500 lbs available forage/acre X 70% (.7) = 2450 lbs grazed forage/acre.

Step 5: Calculate total grazed forage:
2450 lbs grazed forage/acre X 80 acres = 196,000 lbs forage expected from pasture.

Step 6: Calculate deficit or surplus:
Total supply - Total demand: 196,000 lbs available - 316,800 lbs needed = -120,800 lbs deficit.

While they are disappointed in not achieving their goal of year-around grazing, they know at the onset of winter what they are facing and can start making adjustments at the beginning of winter rather than a few months into the winter when their options might be much more limited.

What are some of the options they might consider?

1) They could plan to cull more cows to reduce the demand level. 2) They could try to rent crop residue acres or other underutilized pasture in the area to increase the supply. 3) The inventory estimates were based on three-day stockpile allocation and they could increase intensity to one-day allocation and pick up another 10% grazing efficiency. 4) They could buy 60 tons of hay.

The main point is they have several options to consider while they still have time to make the adjustments without being under extreme pressure.

What if your operation isn't this simple? Let's take it a step farther and look at a diversified operation with spring calving cows, weaned calves to be carried over to graze as yearlings, as well as a spring lambing ewe enterprise.

Once again, we could use any of the three systems outlined in the previous examples. If you like the simplicity of the CDA system, you could convert all of the weaned calves into cow-equivalents and say a weaned calf is equal to .6 CDA. Depending on relative weight, six ewes might equal one cow so a ewe becomes .16 CDA. You can still maintain your pasture inventory strictly on a CDA basis, but make allocations for the other classes of animals based on your personalized conversion factors.

If you're going to go through that much trouble, why not just use standard animal units? Once again, if you work with public agencies, they will want or require you to use standard AUD or AUM format.

One problem that is beginning to emerge in the world of standard animal units is the standard no longer seems to be standard among everyone involved. Some more recent publications use an 1100 lb cow as the standard and factor in a slightly higher intake rate. Whereas the historic standard for one AUM was 780 lbs of forage, some more recent range publications give the value at 915 lbs. If you are strictly working on your own, use an accounting system with which you are comfortable. At least your cow-day or ewe-day relates to a real world creature you are familiar with on a daily basis.

Let's look at this mixed herd and do some calculations. I like putting all the information into a spreadsheet and working from there. If you don't use spreadsheets, just make a basic table that looks something like the one below.

Animal class	# of head	Cow factor	Cow equivalent	Calendar grazing days	Winter requirement as cow days	% of total
Dry pregnant cow	100	1	100	120	12000	54%
Weaned calves	92	0.6	55.2	120	6624	30%
Dry pregnant ewes	200	0.16	32	100	3200	14%
Bulls	3	1.4	4.2	90	378	2%
Rams	4	0.2	0.8	90	72	0%
Total demand:		as CD:	192		22274	100%

In this example, cows and calves are shown to graze on stockpiled forage longer than the ewes or the sire groups. This is simply to illustrate that you might choose to treat groups differently and you need to make adjustments in your calculations. I like looking at the percentage of the total forage resource going to each class of animals and compare that to the percent contribution that group makes to total income or profit share.

In this scenario, the cow herd is taking over 50% of the winter forage. Are they contributing at least 50% of the profitability in your operation? If they are not, maybe you should consider a different centerpiece enterprise.

Now comes the question of whether to graze all these critters as a single herd or split them into several herds. This is not a simple question and there are many factors to consider. Cattle and sheep have different fencing requirements. Is all of the property capable of securely holding sheep? Are some areas of the property more prone to predator problems? Sheep have much lower daily water requirements than do cattle. Can the ewes be put in pastures where there may not be enough water for the cow herd? How much weight do the calves need to put on over winter? Can they achieve this weight gain grazing the same forage as the mature cows? How effectively could you manage a leader-follower grazing scheme? How much time do you have to deal with several different groups? All of these are considerations in your planning.

It is far better to have all of these things figured out in October than to be wrestling with them in February. Having a sound plan based on real information is critical to success.

Planning daily forage allocations:

Thus far we have been planning for the entire winter grazing operation. We've grown our forage. We have our budgeting plan. Now what do we do when the stock are at the gate and we're ready to set up some fence?

Grazing management is really a learn-by-doing process. We can read books and attend grazing schools and seminars, but eventually we have to go out to the pasture and do it for ourselves. One of the really nice things about grazing dry, pregnant stock on dormant pastures is it is really hard to screw it up. That is, as long as you're paying attention to the outcome of your actions.

A good reason to do no more than one-day allocations when you're first starting out is you get to see the results of your actions within 24 hours. If the ground is stripped bare and the cows are bawling, you guessed wrong. Today you give them more. Keep increasing the allocation until the pasture residual is where you want it to be and the cows are content.

If you gave them too much, leave them there another day and make the adjustment tomorrow. As long as you are monitoring cow body condition, observing their behavior, and watching the post-grazing residual, you're probably going to do alright.

For those who feel the need to make those calculations, it's just like what was shown earlier in the chapter except rather than doing it for the entire season, we just do it for one day. In the 6-step calculation, just drop out steps 4 and 6, change 120 days to one day, and you have the calculation for how many acres or what fraction of an acre you need to allocate each day.

The next two chapters will discuss setting up your winter grazing cell to make both fence moving and accounting easy.

Chapter 22
Winter stock water

After the air we breathe, water is the first limiting nutrient for animal performance, even for life itself, yet we often don't give it the attention it deserves. We give water a little more thought in the summertime when it is hot and we know they need a lot of water. It seems one day without adequate water in the hot season makes cattle draw up and look like walking death. In the winter, it is not so easy to see and we're more prone to think they need a little more protein or maybe a molasses tub. It's not until the water freezes up and there is no water in the tank that we start thinking about it and wondering why we didn't do something differently.

As with feed budgeting, a good place to begin planning your stock water supply system is knowing what the livestock will require during the winter months. A dry pregnant beef cow will consume about 7-10 gallons of water when the temperature is below about 55 degrees F. Above that temperature, water consumption increases at a rate of about four gallons per 10 degrees F increase. The base determinant of water required is weight of the animal. Because animals eat in proportion to their body mass and water is required for rumen processing, bigger animals require more water. Dry, pregnant dairy cows have a slightly higher base water requirement per unit of body weight than do their beef counterparts because basal metabolism turns a little faster in high milk-producing animals.

Sheep and goats, as well as elk and deer, have lower basal water requirements than do cattle. Water requirements for sheep are typically 60-70% that of cattle when compared on basis of comparable body weight. That means a flock of sheep with a total body weight of 100,000 lbs will use only 60-70% of the water consumed by 100,000 lbs of cattle, in comparable physiological state of production and body condition. Goats have slightly lower requirements while elk and deer are slightly less than goats.

Once lactation begins, water demand goes up due to the high water content of milk. For cattle, the increase is about three gallons of water per gallon of milk while for sheep or goats it is slightly less. This relationship becomes important when dealing with fall calving cows or fall lambing ewes. A moderate milking beef cow calving in October will reach peak lactation in late December or January. If she is giving 20 lbs of milk at peak, she will need an additional 60 lbs of water or about seven gallons more. It is much more difficult for a lactating cow to rely solely on snow as her water source than it is for a dry cow. The challenge is doubly great for dairy cows in winter grazing systems.

Water requirement for growing stock is directly tied to rate of gain because higher rates of gain require more forage intake and more forage intake requires more water for rumen processing. For beef cattle, the average water requirement is roughly one gallon per 100 lbs of body weight. Because they tend to be more selective grazers and have less winter experience, yearling cattle also have a greater difficulty meeting all of their water needs just from snow.

If you have a free flowing water supply, knowing the demand rate helps you calculate the necessary pipe and tank sizes. If your water system was designed for summer demand, you are pretty well assured it will be adequate for winter supply, as long as it keeps flowing. If you plan to haul water, knowing the daily requirement is essential to ensuring your stock get adequate water.

Snow as a water source:

More and more graziers are finding they can use snow as the primary stock water source for winter grazing. Several research studies conducted in Alberta compared performance of dry, pregnant beef cows; cow-calf pairs; and weaned calves with either just snow available or heated water tanks. They found no significant performance differences of dry cows or pairs between the two water sources. Weaned calves performed better with continuously available heated water. Many Canadian cattlemen rely solely on snow for their winter water. The Canadian researchers emphasize eating snow is a learned behavior and close monitoring of cows on snow is essential.

My first experience using snow came on our Missouri farm before we had all of our pipeline system and tanks fully installed. We had cows in a stockpiled pasture where the pipeline ended. There was no permanent tank installed, so I had to fill movable water tanks morning and evening. We had used these tanks for a couple of weeks so I had a good idea of how much water the cows were drinking every day. Then it snowed.

I went out in the morning, filled the tanks the cows had drunk down the evening before, and went to work at FSRC. When I came home that evening, the tanks were still full with a couple inches of ice on them. I assumed they must have frozen over before the cows got back to them for a drink in the morning. I broke the ice and called the cows. A few wandered over, looked at me, and then went back to where they were grazing. For the next few days I went to fill the tanks but they were staying full. I decided the cows were doing fine on snow and from that time forward, if there was at least a couple of inches of snow on the ground, we didn't bother watering them.

We had a similar experience in Idaho grazing stockpiled pasture the first winter we had the new stock water tanks and pipelines in place. That first winter, I carelessly let some tanks and lines freeze that I should not have, so by early December we were reduced to using just one tank out of the six in the system. The cows had to walk about 3/4 mile back to that tank.

We were using three-day strips that winter and I noted some interesting watering behavior. For the first two days on a strip, the cows did not come back to the water tank at all. Mid-morning on the third day, they all came to drink at the tank. This pattern repeated itself for several weeks.

We had been getting a dusting of fresh snow almost every night so it was very easy to look at the tracks to see what was going on. On a new strip, there was plenty of fresh snow for them the first day and into the second day. Every time they took a bite of stockpiled forage they were also taking in a substantial amount of snow. By the third day, all the easy to consume snow had been utilized and the remaining snow was hard packed and unavailable to the cows.

There remains a debate among graziers whether cows can actually lick up enough snow to meet their needs or do they only take in snow as a bonus as they graze. I do not know the answer to this question, but here is what I have observed.

1) I have never seen a cow lick just snow. 2) We sometimes have dry powder snow and sometimes we have wet heavy snow. Cows are much more likely to use the stock tanks when we have powder compared to wet snow. 3) Cows use the stock tanks much more when we are swath grazing than when we are using standing stockpiled pasture. 4) Cows on hay are at the tanks almost as often during the day as in the summertime.

These observations lead me to believe that the primary avenue for snow consumption is as a byproduct of grazing standing forage. When the snow is powdery, it sifts out of the bite and they consume little. When it is wet, every grazing bite takes in substantial snow. With swath grazing, they are only eating from about 10-20% of the surface area of the pasture so potential water intake is greatly restricted. Snow between swaths is frequently undisturbed. When eating hay, they consume very little moisture from the hay so they are totally dependent on the tanks. Perhaps other cows in other environments do a better job of just eating snow, these don't.

My bottom line is you need to closely observe what

cows are doing and monitor their appearance and behavior when using snow as your primary water source.

Providing winter water:

Flowing water is one of the best ways to keep stock water available. In my experience, the ideal winter water system is a pipeline system from a spring development with enough elevation drop to allow gravity to do the work. Gravity is one more free resource we should try to utilize whenever it is feasible.

If there is enough fall, design a stock water system with a series of cascading tanks and an overflow drain out of the lowest tank. With a series of cascading tanks, the overflow pipe from the first tank becomes the supply line for the second tank and so on. This setup allows water to flow through all of the tanks, thereby keeping the entire system open with the need for only a single dump point or drainage field. If each tank is on a separate line, each tank requires its own overflow and drain field. This will take a lot more water flow to keep multiple tanks open and the added expense and/or inconvenience of multiple drain fields. With 40 to 50 ft of drop between each successive tank, gravity flow will develop about 20 psi pressure at each tank. With appropriate pipe and valve sizing, this system can water even large numbers of head.

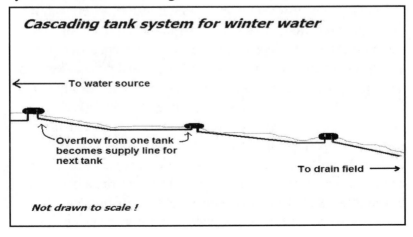

Cascading tank system for winter water

To water source

Overflow from one tank becomes supply line for next tank

To drain field

Not drawn to scale !

On Circle Pi Ranch where we live, our winter water system starts with a spring about 100 ft higher than the pastures. Where we are grazing stockpiled forage on a center pivot, there are six stock tanks equally spaced around the inner circle fence located halfway between the pivot center and the outer reach. During the summer months we run the stock water system off the irrigation mainline. At the pivot center is a 6-valve manifold with a shut-off valve for each of the six tanks. There is also a control valve to send water from the mainline into the manifold and then out to each of the six tanks. When winter comes we shut off the mainline supply valve and open a valve from the spring.

Water flows from the spring and joins the stock water system at a "T" near tank # 1. It flows on to the pivot center where it supplies all six tanks through the valve manifold. Each tank is set up with a bleeder valve to keep the water lines open and when the cattle come to a new tank, we can switch from the bleeder valve to the water flow valve and have fresh water to the tank. We then use an overflow to keep the current tank open.

The week I wrote this particular chapter in December of 2009 we experienced double-digit sub-zero lows for several days with highs climbing to the low single digits. Never once did the stock tank in use freeze over. A well-designed spring fed stock water system can make winter watering very simple.

Even a productive well or large pond with pumped water can be used effectively in a cascade system as it conserves the amount of water that must be overflowed to keep the system open. Obviously mild climates require less water flow to prevent freezing than do severe winter environments.

There are commercially available stock water tank valves that operate with a thermocouple-controlled secondary valve that increases water flow when the tank water drops to a certain temperature. These units are very useful in areas with moderate winter temperatures or widely fluctuating temperatures as they restrict overflow water only at times when the likelihood of freezing is greatest. These units can be used in conjunction with virtually any type of pipeline and tank water system.

While a free flowing stream can also provide a viable winter water system, I recommend using a limited access approach to prevent livestock from damaging streambanks or polluting the water with feces or mud from their hooves. In some parts of the USA and Canada stream usage by livestock is an increasingly controversial topic. We should be proactive in our management to minimize negative impact and demonstrate our concern for clean surface waters and healthy aquatic systems.

If continuously flowing water is not an option, my next preference is use of some sort of energy-free, freeze-resistant waterer. I do not like the term freeze-proof waterer as any piece of equipment can and will freeze if improperly installed or maintained. Most of these systems work well down to about 0 degrees F and then colder temperatures start sorting out which are better than others.

In my experience, the floating ball type waterers (e.g. Mirafount and Ritchie) will work well if installed perfectly level on a pad and provided with a windbreak on the two sides where prevailing weather strikes, usually the north and west in much of the USA and Canada. Where I have seen them fail is when the unit is out of level and the ball(s) on the high side do not seal properly. Wind sweeps in around the ball and freezes the water inside. Without windbreaks in extreme cold, we have seen just the water dripping from the animal's muzzle collect in the groove where the ball meets the tank body and almost immediately freeze the balls in place. This is much more of a problem with sheep or calves than it is with mature cows or large yearlings. Attention to installation and using windbreaks minimize the problems with these units.

The deep tube-type waterers that utilize earth heat rising in the tubes to keep them open also work reasonably well when properly installed (e.g. Cobett or Thermo-King). As with the ball type waterers, these require a certain number of animals using them regularly to keep them open. More frequent use results in greater rate of water exchange and helps prevent freezing. These units also benefit from windbreaks installed on the prevailing windward sides. Whereas the ball type waterers perform best with a concrete pad, the tube waterers do very well with just a geo-textile and aggregate pad around them to help with drainage and livestock support.

I have seen both the ball-type and tube-type waterers perform well even in climates where -30 degrees to -40 degrees F temperatures are regular occurrences as long as they are properly installed with windbreaks. These installations are fairly expensive compared to open stock tanks. Remember, you do not need to have water as readily available to stock in the winter as they need it in the summertime. If only 15-20% of the water installations in your grazing cells are usable in the winter, that is probably good enough.

Another useful device that can be used in stock tanks or ponds for keeping water open is a solar powered DC air compressor. These units have an internal rechargeable battery to continuously power a small air compressor. The constant flow of bubbles can keep an 18- to 24-inch opening in the ice down to -10 to -20 degrees F. Even at those temperatures, the morning ice layer is usually thin and easily broken by cattle. Sheep may need a little help.

These units can make any open stock tank into a year-around water source. The best way to install the air line is with a "T" in the line so two columns of bubbles can be maintained. Place one end near the water supply pipe and float valve. This will keep the float working even in freezing conditions. Place the other end near the tank edge to maintain an open drinking space. Use a weight to hold the line near the bottom of the tank. If the line floats to the top, it will just release air to the atmosphere and the tank will freeze.

Heated waterers, whether they be electric or propane tend to be the most expensive winter water option. Initial installation cost may be similar or even lower than energy free units, but the long term energy costs make them more expensive in the long run. They are a last resort in my opinion. We had no heated waterers on our Missouri farm, nor do we use them in Idaho.

Regardless of the water system being used, we generally begin winter strip grazing close to a water source and work away from it. This strategy works particularly well in areas where the ground freezes and remains frozen for extended periods of time. Stock going back and forth across frozen ground do little damage to the pasture plants there. If you are in an area where frequent nighttime-daytime freezing and thawing occurs or warm periods (highs > 40 degrees F) can occur during any winter month, allowing the stock to travel back and forth across previously grazed strips is much less desirable as trampling damage can occur.

Preferably, you will have multiple water locations so stock are not in the same area longer than two to three weeks and then the pastures should be empty for the balance of the dormant season. When we were on our farm in Missouri, cattle were rarely on any paddock longer than a week because we could provide water in many different locations. Mud is a much greater challenge to year-around grazing than snow cover in most areas south of the Iowa-Missouri boundary.

It's a good idea to visit other graziers who have been grazing deep into winter for a number of years. Most of them have already been through the trial-and-error process of figuring what works and what doesn't work in your environment. Try not to reinvent the wheel. It's a lot more cost effective to learn from someone with experience.

Chapter 23
Fence systems for winter grazing

There are two basic approaches to grazing cell design or infrastructure: fixed or flexible.

Fixed designs use permanent fences and stationary watering points while flexible systems use temporary fences and movable water tanks within a framework of permanent fence. There are also variations between the two, the most common of which is using permanent watering points, but movable fences.

The systems we operated on both our Missouri farm and on the ranch in Idaho are combination systems. Fully portable water systems are much more difficult to use in the winter compared to the growing season, so combination systems make a lot of sense for year-around grazing.

Fixed systems work well on large operations due to the lower installation cost per acre for large grazing units compared to smaller grazing units. It takes fewer feet of fence per unit of area enclosed for increasingly larger areas.

For example, let's consider the fence required to enclose three perfectly square paddocks 1, 10, or 100 acres in size. Each of these paddocks has a perimeter distance of 835, 2640, and 8348 ft, respectively. The corresponding distance of fence required to enclose an acre is 835, 264, and 83 ft, respectively. The cost per foot of fence is also lower on larger units as each stretch of fence becomes longer, thus the investment for fencing a small pasture is much higher per acre than it is to do a large pasture.

It is animal product that ultimately has to pay for all pasture investments, so permanent fencing is much more affordable on large properties. Smaller properties need to minimize their investment in infrastructure, so flexible systems are a better choice on smaller properties.

We can look at stock water development with the same logic. A winter stock water installation may cost from several hundred dollars to well over a thousand dollars depending on size and type. One tire tank at the center of 40 acres serving four 10-acre paddocks costs four times as much per acre than the same tire installed at the center of a 160-acre pasture serving four 40-acre paddocks. There is a definite economy of scale even in grazing systems and it definitely favors the big guys.

Grazing cell layout for efficient winter grazing

When it comes to winter grazing, you need to have a lot of flexibility in the feed allocations you provide. Our Missouri farm was 260 acres and was subdivided into 70-some permanent paddocks, but we did further strip grazing within each stockpiled paddock with polytape. On that farm most of our strip grazing fences were only from 220 to 450 ft long as most of the paddocks were only three to four acres. These short distances meant chore time was very short. Fences of this length can easily be taken down and put back up in just five to ten minutes.

On the next page you will find a simplified illustration of one 80-acre grazing unit we had. This unit had a few patches of woods, a large pond area fenced out, as well as an ephemeral stream that ran about two thirds the length of the property. Even though these subdivision fences were just single wire hi-tensile fences where we just ran cattle or three-wire where we ran sheep, it was probably more infrastructure investment than we really needed.

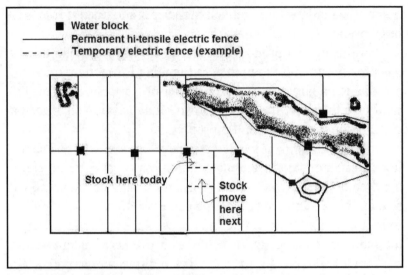

This 80-acre unit could have been managed more effectively with just single fences down the center of the property and along each side of the wooded stream and then all the growing season subdivision and winter strip grazing fences done with polybraid on geared reels and step-in posts. The overall cost of the infrastructure would have been much less and we would have had greater management flexibility. Figure 2 shows an alternative layout with minimal permanent fences.

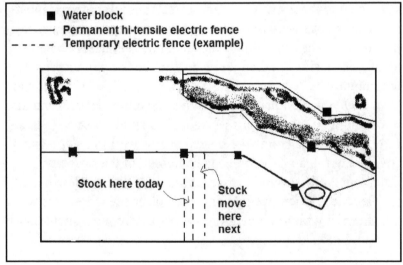

The only thing I don't like about this second layout is the strips are more long and narrow than I really prefer. After the first couple strips on stockpiled pasture it would be fine as the stock would tend to move straight ahead into the strip rather than moving up and down through the strip. In this example, the temporary fence runs would be about 660 ft which is a very comfortable length with which to work.

Another strategy would have been to split the 80 acres into four grazing corridors rather just two as shown in Figure 2. This would shorten the temporary fence distances to about 330 ft and make each allocation less extreme in length:width ratio. Using this layout would require three permanent or semi-permanent fences the length of the property rather than just one (i.e. 1 1/2 miles vs. 1/2 mile). It would also require more pipeline and water points.

You will need to weigh the added infrastructure cost against the increased convenience in your particular situation. There are always tradeoffs.

In Missouri we used polytape for greater visibility for benefit of both the cattle and the bountiful white tail deer population we had. When we tried to use polywire, we experienced much more fence damage from deer. In those days we were using much lighter polywire than what is now available. The heavier braided polywire is much more visible and seems to do as good a job as polytape had done. The main problem using polytape in the winter is it collects ice and snow cover much more than does polywire. If your conductor gets weighted down with snow or ice, it is very quickly on the ground and less than well trained stock will not respect it any longer.

Here in Idaho we do all of our winter subdivision using portable fence within a framework of just two permanent fences on each center pivot, one on the outer reach of the irrigation and the other located halfway between the pivot center and the outer fence. Figure 3 shows a typical center pivot grazing cell layout with potential for using six different stock tanks.

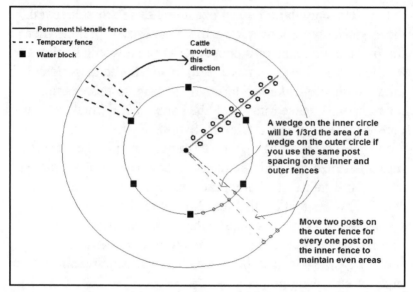

Permanent hi-tensile fence
Temporary fence
Water block

Cattle moving this direction

A wedge on the inner circle will be 1/3rd the area of a wedge on the outer circle if you use the same post spacing on the inner and outer fences

Move two posts on the outer fence for every one post on the inner fence to maintain even areas

The first winter we were here, the only available stock water on this 300-acre pivot was an irrigation ditch which crossed through the pivot to the next ranch. We just left water flowing in the ditch as long as the cows were here. There were no interior fences, just a hi-tensile electric fence around the perimeter. About one-third of the pivot had been seeded to barley-oats-winter pea mixture and that crop provided the basis of most of our grazing that fall. Another 100 acres had a poor stockpile growth on it. We had 428 dry cows to work with that fall.

From the irrigation ditch to the outer fence at the greatest distance was nearly 3000 ft. Most of our strip grazing that winter was made with roughly 1/2-mile or longer runs plus we had to keep extending a fence along the flowing irrigation ditch to provide access to water. It is no fun running those kinds of distances of temporary fence in the summertime and even less fun in the winter.

Why isn't it fun? 1) It requires larger fence reels holding at least a half mile of polywire, which are much more cumbersome to use than the standard reels which hold only about a quarter mile of polybraid. 2) Spacing step-in posts at distances even up to 100 ft still requires more posts than you can comfortably carry on foot. 3) You spend much more time out in the cold doing your chores. 4) When the elk came off the mountain every night, it was a longer stretch of fence to reset.

Because of the less convenient process of setting up a half mile of fence rather than just a few hundred feet, we reverted back to giving three-day strips rather than moving the cattle every day. Our high yielding barley-oat-winter pea mixture was left standing, not swathed, so the waste with three-day grazing allocations was significant, particularly after snow got a little deeper and crusted. We did harvest 292 CDA off that crop from late September through mid-December in spite of the waste and feeding lots of deer and elk.

That first winter gave us a lot of experience on what we didn't want to do. The following summer we were able to install the stock water system and build the inner circle fence. The stock water system should have allowed us to keep the cattle closer to water and minimized the amount of back grazing we allowed. However, I didn't get some winterization taken care of in timely manner and ended up allowing five of the six tanks to freeze up and we were reduced to using just one tank. With the inner circle fence in place, we only had to run 1000 ft stretches of temporary fence rather than 2000 to 3000 ft. Back in Missouri I would have considered a thousand feet of temporary fence to be way too much. Now I consider it normal. My winter chore time for feeding typically 300 to 500 beef cows is 30 to 40 minutes per day.

For those of you thinking about stockpiling and grazing center pivots, remember the times and distances I am citing here are for a 300-acre pivot, not a standard quarter section pivot where the temporary fence distances would only be about 660 ft and the time required for a daily paddock shift would only be 15 to 20 minutes if the pivot were set up similar to what is shown in Figure 3.

If you live in a region that experiences very cold temperatures, the amount of time required to move the stock to a new strip and set up the next move should be kept to what you can comfortably tolerate. There is a lot of individual variation in how well we tolerate the cold. I personally would prefer to spend an hour out on foot in sub-zero weather than mess with starting a tractor and feeding hay. Others may prefer much shorter fence distances and spending less time out moving fences.

To make winter feed budgeting and allocation easy, we like to build the accounting system right into the fence system. By making the grazing corridors as uniform as possible with parallel fences, you can use a predetermined post spacing in the permanent fences so that the strip between each pair of posts is a known acreage. For example, if you make the grazing corridors 435 ft wide and set permanent posts every 50 ft, the area between the parallel fences and each pair of posts would be one-half acre. For those who don't have instant recall, an acre is 43,560 sq ft. You can use any line post spacing you want in combination with different corridor widths in order to make the accounting work.

Planning the fence layout on a center pivot takes a little more calculation, but is actually fairly simple if you keep a couple of basic geometrical relationships in mind. On a center pivot, we still use the same post spacing on the inner and outer circle fences. If the inner circle fence is pretty close to half the distance between the pivot center and outer reach, you will maintain even segments of the pie if you always move two posts on the outer circle for each post on the inner circle.

Again using an example with 50 ft line post spacing, for each 50 ft on the inner circle, go 100 ft on the outer circle. For computing acreage within each pie segment simply take the average of the distance on the inner circle and outer circle. In this example, it would be 75 ft. On our 300-acre pivot, the area in each of these pie segments in the outer circle is about 75,000 sq ft or about 1.7 acres. On the pasture inside the inner circle fence, the pie segments come to the pivot center for an inner point measurement of 0 ft. The average of 50 ft on the inner circle fence and 0 ft at the pivot center is 25 ft, so each inner circle pie segment is 25,000 sq ft or 0.57 acres. To give the same size allocation on the inside circle we would move three posts. Another way to remember it is an inner pie piece will always be about 1/3 the acreage of the corresponding outer pie segment. Figure 3 has a graphic illustration of this principle.

You probably always wondered where high school geometry would finally reappear in your life.

Choosing fence materials that work in winter

There are three main components of a temporary fence unit: 1) the reel, 2) the conductor, and 3) the posts. Here are the features I think are important for each component used in a winter grazing program.

I like a geared reel rather than straight-crank reel. Why? Because it shortens the length of time it takes to retrieve a fence. When it's 20-below, this becomes an important consideration. Some graziers like to use a power winder to retrieve their polywire. This may be okay on a straight-crank reel, but it will void the warranty on most geared reels. Power winding polywire rather than walking while winding the spool also reduces the life expectancy of most polywire by about 50%.

The plastic spool should have some flexibility in it. Rigid plastic tends to get brittle in cold temperatures and is much more prone to breaking even with normal use. I remember buying a couple cases of cheap reels when we lived in Missouri. A real bargain at $12 apiece. Not one of the eight reels made it through the first winter without at least part of the spool breaking. Cheap is not necessarily a good thing.

A reel with a wire guide is much easier to use than one without a guide, particularly if your reels are filled to the limit. One problem with a geared reel without a guide is you can wrap up a lot more polywire around the crank with just a few turns of the handle than you will get with a straight-crank reel. When it is extremely cold outside at chore time, I prefer to not have unexpected problems with my equipment.

I have found some reels are much more ergonomically comfortable than others. A reel that you have to hold out in front of you rather than suspending below your hand is much more strenuous to use. This becomes particularly apparent in cold weather when the stiffness in your wrist and hand are aggravated by the cold.

My personal preference is the O'Brien 3:1 geared reel or those brands built off the O'Brien patent. We have been using these reels since the late 1980s and have found them the most satisfactory all-around electric fence reels.

We are using braided polywire almost exclusively for all of our temporary fence applications these days. When we first arrived in Idaho, we used some polytape to get the attention of our resident antelope herds. When they were tearing around at top speed, they just weren't seeing the lighter polywires. We trained them with the polytape and then when PowerFlex polybraid became available in 2005 we switched to this product and have been very pleased with the conductivity, durability, and visibility. We no longer have any issues with elk, deer, or antelope with our temporary fences.

Braided polywire has greater conductivity than typical twisted polywires because the increased surface area of having bundles of three conductors braided together and then those bundles braided with the plastic filament bundles. Breakage of the fine conductor wires is lessened with the increased strength of the cable effect. If one bundle breaks, the two pieces remain in contact with other bundles throughout the length of the fence and overall conductivity is only minimally affected.

The polybraid is much larger in diameter than most common polywires so some standard fence reels designed to hold a quarter mile of twisted polywire won't accommodate a full quarter mile of polybraid. The larger diameter of the braided product and two-color plastic create a very visible temporary fence.

We have watched from the house as groups of 20 to 40 elk running across the pivot come to a complete stop for a single strand of polybraid and then one at a time jump the fence. These are the same animals that will barrel headlong into a five-strand barb wire fence and pass right through leaving a wake of destruction in their path.

I will emphasize we never turn off any sections of permanent electric fence even if we don't expect cattle to be in that area again for seven or eight months. We never leave a section of temporary fence non-energized. This is how we maintain the respect of the wildlife for our fences and keep maintenance to a minimum. Five miles down the valley there are miles of three-strand hi-tensile fence strewn across the desert by elk and antelope because the managers took the energizer away when the cattle left the range. For two-thirds of the year, the wildlife were allowed to run freely through the non-energized wires and they never learned to maintain respect. Of course the local point of view is electric fence won't work here because the wildlife will trash it when the reality is it was lack of management that trashed the fence.

The final component of an easy-to-use and effective winter fence system is the portable post. There are lots of options available and most graziers have developed their own preference over years of use. Winter grazing puts additional demands on a portable post including the difficulty of getting a post in frozen ground, cold-induced brittleness, and ice and snow load on the fence.

I prefer to not carry any unnecessary tools around with me when I am moving fences, particularly in the wintertime, so I look for features in a step-in post that will allow me to use them effectively in the cold. Characteristics I look at are the diameter, length, and strength of the spike; size of the step relative to my foot; flexibility in the cold; and basic durability.

Many of the step-in posts have a 5/16" or 3/8" diameter spike from 5" to 7" in length. Anything that size is going to be difficult to put into frozen ground without a pilot hole. A pilot hole requires a tool of some sort, which is one more thing to carry. In order to be able to step a post into the ground, you need to be able to get and keep your foot on the step. Some posts have steps only an inch or two in length. If you have snow or ice caked on your boot, it is very difficult to deliver enough force to get the post in the ground. If you are making a pilot hole, this is much less of an issue. Some plastics become very brittle in cold weather and as you try to force the post into frozen ground they simply snap in your hands. Pretty aggravating.

We have been using the O'Brien Treadline Step-in for over 20 years now and have found them to be the most satisfactory of the plastic step-ins. The spike on these posts is a 3/16" hi-tensile rod only about 4" in length. The step is a full 4" wide to accommodate most of your boot width and they maintain better flexibility even in very cold conditions. We usually use about 100 to 150 posts during our winter grazing period and will expect to break about five posts each year. This usually occurs when overflow water from the tanks drains out to where we have a fence set up and the ice locks it into the ground. I have learned to just leave those posts out there and recover them in the spring.

A 5% annual breakage rate means the average life expectancy of a box of posts is about ten years. When we moved from Missouri to Idaho in 2004, I brought 100 O'Brien step-in posts with us. I had last purchased a new box of posts in 1994, the year Dawn sold her first fence company, Green Hills Grazing Systems, to Dave and Connie Krider, current owners of PowerFlex Fence Systems in Missouri. At the time of writing this book, my newest O'Brien posts were 15 years old. I can still bend a 15-year-old post over into a complete U and it will spring back to vertical. That is durability.

Some graziers prefer the steel pigtail posts, especially for winter use. The all-steel construction does allow you to deliver more force on the post while shoving it in compared to most plastic posts, and a hammer can be used on the step to assist with driving when needed. Although I don't use pigtail posts myself, I recognize it is much easier to carry a dozen of them in your hand compared to the O'Brien step-ins. If that's what works for you, go with it.

Another thing to keep in mind when doing tight strip grazing with daily or more frequent moves is the post spike doesn't have to go all the way into the ground. When it is really cold, I'm satisfied with getting the post just an inch into the frozen ground knowing that it only has to be up for 48 hours. Before we had our wildlife so well trained I worked a lot harder to make sure posts were deeper in the ground. If you experience high winds, you will need to make sure the posts are in deeper. Another advantage to our dry climate is the snow is usually just powder and doesn't accumulate on the polywire like a heavy, wet Midwest snow wants to do. A lot of times in Missouri when we had one of those heavy snows, I just stuck the posts in the snow and didn't even worry if it penetrated the soil at all.

There are a number of ways to deal with frozen ground when you finally decide you need to do something different than just trying to step the post into the ground. Pilot holes are one option and can be created several different ways. One common practice is using a cordless drill to make a hole for the spike. Just use a bit the same size or slightly smaller than the post spike diameter to keep the post from getting sloppy in the ground. Some back-to-basics graziers use a hand brace rather than the cordless drill to avoid the disappointment of a dead battery half way across the paddock.

You can also punch a hole with a spike welded to the bottom of a pipe or rod. For you Westerners, a 5/16 jack fence spike welded on a 3 ft X half-inch rod makes a nice pilot hole for the larger diameter step-in spikes. For the O'Brien posts or others with similar 3/16" diameter, use a spike from a broken post welded to a half-inch rod for your punch. This can either be jabbed or pounded into the ground. If I ever have to resort to this, I will make the pilot hole punch as a slide hammer so I only need to carry one extra tool, not two.

Some graziers like to make a portable post base for step-ins. I have seen everything from concrete circles to just a piece of 2x8 with a hole drilled in it. At FSRC we made some concrete post bases by slicing five gallon buckets into two-inch-wide plastic bands and using these for little mini-forms for concrete. A piece of Premier SupaTube set vertically in the center perfectly accommodates the 3/16" spike of an O'Brien step-in. For larger diameter spikes just use a piece of 5/16" or 3/8" plastic tubing.

Each of the concrete circles had a single cattle panel square set in it vertically to make an easy handle for picking up and moving the base. If you have a fence with ten posts, you need ten of these bases and then you are pretty well committed to using an ATV or other mode of transportation for every fence move. Using a 12" piece of 2x8 with a hole drilled in the center is lighter than hauling around concrete rings, but still becomes cumbersome to carry around without wheels.

What do I do when I just can't get a step-in post in the ground? I've actually never had the experience. There have been numerous times when I was within a hair's-breadth of throwing in the towel, but the O'Brien step-in has always prevailed for me. I do note a huge difference between the difficulty of getting a post in the ground at -20 degrees F compared to 0 degrees F compared to +20 degrees F. At double-digit sub-zero temperatures, it takes me about ten minutes more to move the 1000 ft of fence due to greater challenges with the posts.

Moving temporary fence in the winter time can either be easy or hard, it all depends on making the right choice in fence components.

Chapter 24
Swath grazing

In the Eastern half of the USA most winter grazing is done on stockpiled pasture or standing winter annuals. This is just accumulated growth left standing in the field after the growing season ends. Stockpiling works very well for many perennial grasses and some legumes allowing livestock to graze this material throughout the winter. Winter annuals are more susceptible to damage from freezing temperatures and sometimes the window of opportunity we have in which to use those crops just isn't long enough. That's part of the reason some people end up resorting to making hay. Put it in a bale to preserve the quality and make sure you have feed available. The problem is it is really expensive.

What if it you could just cut it down at that point and leave it in the windrow to graze later? It would save the cost of baling, hauling, and feeding hay and possibly manure hauling if the livestock were dry lotted for winter feeding. That is what swath grazing does. With crops like alfalfa or winter annuals that are highly susceptible to freezing damage, swathing just ahead of or right at the time of the first killing frosts locks in the quality at that particular growth point. Once the forage is in a windrow in a dry climate there is little change in forage quality and only minimal dry matter yield loss.

Besides preserving forage yield and quality, swaths or windrows can be accessed by livestock through much deeper snow than typical standing stockpile. The larger the windrow, the more snow with which livestock can contend while maintaining body condition. In some situations it pays to rake together two or more swaths to form a bigger, boxier windrow. Last crop alfalfa swaths frequently are not large enough to work well without raking. Good crops of annual forages can make large windrows from just a single pass of the swather or mower-conditioner.

By just putting the forage in the swath and grazing it directly, the cost savings is typically between $25 - $35 per ton of forage fed. The labor cost for grazing is much less than feeding hay. The more forward planning you put into designing your winter grazing cell, the more time and money you will save.

Where does swath grazing work?

Unfortunately swath grazing just doesn't work in more temperate, wetter climates. The forage rots in the windrow before it can be utilized. In drier climates, though, windrows can lay there for many months with little loss of yield or quality. I know several ranches where cows graze windrows December through March.

Figure 24.1 shows the parts of the USA where swath grazing may work. In addition almost all of the Prairie Provinces of Canada are prime swath grazing country. Alberta and Saskatchewan are where swath grazing was really developed and most of the kinks worked out.

Swath grazing has the greatest opportunities in the Intermountain West and the Northern Plains. These are areas where it is dry enough to have very little risk of feed deterioration once it has been put into the swath. Farther down in the Southwest states, winter rainfall can limit the use of swath grazing. It all depends on local precipitation patterns in those areas. Irrigated land where a crop can be grown late in the season and then the water shut off is the ideal situation for swath grazing.

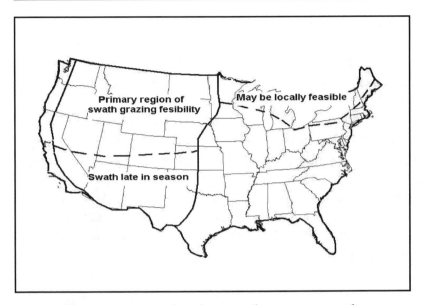

There is a narrow band across the extreme northern states from Minnesota to Maine where early snows can cover swaths and protect them from further weather deterioration. The problem is these are the same areas where snow can get ten feet deep and shut down all livestock movement. Swath grazing may work as a month or so extension of the normal grazing season before snow gets too deep in these areas.

What to swath

In the Intermountain region and on the Northern Plains, annual small grains and legumes are often planted for hay or grazing. One of the problems we have encountered is the small grains getting more mature than we would like them to be at the end of the growing season. Ideally, I like to have a small grain just at the boot stage when it goes into the swath. If you are planning for just a single crop, delaying planting until 60 to 75 days before the end of your growing season will have the crop at the proper stage for swathing about the time killing frosts hit. We consider our pasture growing season done by mid-September, so our target seeding date is between July 1 - 15.

Some people might ask, "Don't you want the grain

component in the swath?" What you gain in additional energy by letting grain form doesn't offset what you give up in stem and leaf digestibility and protein content. Total dry matter yield is greater with a mature crop, but protein and digestible energy yield per acre are greatest at the boot stage.

Another common question is why not just swath the crop earlier in the season whenever it reaches the growth stage you want? The problem with this approach is there may still be too much hot, dry weather before winter and the swathed crop can become very brittle and begin to deteriorate. Hot and dry is much harder on maintaining forage integrity than is cold and dry. Farther north and at higher elevations early swathing can work as there is less heat to make the crop brittle.

With some crops you can take an early grazing and then swath the regrowth. This is when it pays to use Italian ryegrass or a forage variety of triticale, oats, or barley to ensure enough regrowth to justify the cost of swathing. A strictly grain cultivar of any of these crops may not regrow if grazed after jointing has begun. Italian ryegrass is an interesting plant as it can be planted in the spring and it will not head until it has passed through a winter. The likelihood of Italian ryegrass regrowing enough in the latter part of the season to ensure a nice crop for swathing is fairly high as long as you can supply late season water and it has adequate N available.

Winter peas are another example of a valuable crop that rapidly turns into nothing with hard freezes if left standing, but put into a swath with ryegrass or a small grain mixture and it becomes valuable winter forage. The majority of farmers and ranchers don't have a real appreciation for the yield and quality potential of winter peas for swath grazing because of the perceived high seed price. Rather than planting the recommended 15 to 20 lbs/acre in a small grain mixture, most graziers only put 5 to 10 lbs in the mix. Even if peas cost a dollar a pound, seeding another ten pounds per acre can easily increase yield by a half ton per acre. Once again, the extra feed cost only $10/ton, making it some of the cheapest feed you will ever buy.

Last crop alfalfa is increasingly being put into swaths for later winter use. In the Intermountain region where we live, the normal production practice for alfalfa fields is two cuttings of hay and then a light aftermath growth used for grazing. That can easily be converted into a couple of early grazing cycles and then 45 to 60 days of late summer growth to be put into a windrow. The two winters that we swath grazed alfalfa into January really made me a believer in this strategy as an alternative to putting alfalfa in a bale.

Swathing grassy hay meadows

It seems to be an increasingly popular practice to swath grass-dominant hay meadows in the Intermountain region and Northern Plains. These are often swathed at midsummer when the reliable spring rains are past and then left until fall and winter for grazing use. Several research studies from Nebraska to Alberta have shown swaths can preserve forage quality better than leaving the field stand as stockpiled feed. The question becomes is the yield and quality preservation great enough to justify the cost of swathing?

Extended summer heat is the enemy of forage quality in these stockpiled meadows. The longer summer heat lingers past the end of the main growing season, the greater the value of putting grassy meadows into a swath. In the more northern part of this region, the greater benefit of swath grazing comes from making the forage more available to the animals in deep snow rather than preserving yield and quality.

This is a good place to maintain flexibility in your utilization plans. In growing seasons with more precipitation than usual, forage quality stays higher and new growth may occur quite a bit later in summer than typical years. Leaving the forage standing for winter stockpile use may be the more cost effective option. While in a drier year putting it in a swath may be the better choice. The hotter, drier weather might result in quite a bit of deterioration in the swath but it will probably be less than what would have occurred if it had been left standing.

Feeding management

As with every other type of winter grazing discussed in previous chapters, tight strip grazing maximizes utilization efficiency. One Nebraska study evaluated forage wastage with different grazing strategies. When fences were moved every 10 to 14 days, the forage waste was over 25% while every-day moves resulted in less than 5% waste through the winter grazing season.

My own observations between one-day and three-day strips suggest substantially more waste with the three-day allocations. This came about when I had traveled away from the ranch on extended speaking or consultation trips and I had set up fences for less frequent moves to make it easier on Dawn while I was away. When I came back home, the increased waste was visibly apparent. More forage was bedded on and fouled with manure. With a daily move made in the morning, there is very little left by evening for them to bed on.

Orienting your strip grazing fences perpendicular to the swaths is a good idea when you have snow cover. The exposed end of the swath just beyond today's fence provides a good starting point for grazing in the new strip. Cattle and sheep will work their way along the face of the swath breaking deeper snow as they go.

One of the things we learned about using step-in posts when swath grazing is to always put your post through a swath rather than between adjacent swaths. The insulating effect of the swath keeps the ground a little softer and makes the step-in much easier to put in place.

Chapter 25
Supplementing stockpiled pasture

The goal of strategic supplementation is to supply a nutrient or nutrients lacking in the base diet. In winter grazing situations, the lacking component is most likely to be protein, although some stockpiled forages or crop residues may have inadequate energy even for maintenance requirements. We will focus primarily on protein supplementation.

Before we get deeper into supplementation, there is another very important nutritional concept that has bearing on the need for supplementation. I'll call it the grazing layer effect.

We often look at a pasture and think in terms of, is this cow feed or calf feed? Or is this maintenance or growing feed? A better way of looking at the nutritive value of a pasture is how much of it is calf feed and how much is cow feed. I think we get hung up on an all-or-nothing perspective of a particular pasture. We think we have to use it either just for cows or just for calves. The truth is we can allocate it in layers to different classes of livestock using leader-follower grazing.

Almost every forage resource has some growing quality feed, some maintenance quality feed, and some that is just garbage. It is what I have been calling the good, the bad, and the ugly earlier in this book. With managed grazing we can let calves in to harvest just 20% of the forage and then move them on. What is left is still likely to be adequate dry cow feed.

If we had made the young stock use all of that pasture,

we probably would have needed to supplement them for the entire period they were out there. By letting them cream the best forage off the top, we eliminated the additional cost of supplementing. The cows will probably do okay on what was left behind. If they do need supplementation, we can use a lower cost option such as non-protein N (NPN) for mature animals that doesn't work as well for younger stock.

Just keep in mind you can fine tune your grazing utilization to minimize the need for supplementation.

Minerals

Mineral supplementation is a very local and personal question as the need for mineral supplements depends on existing soil conditions, plant community, past management history, current grazing management, and specific performance goals. If your herd or flock is short on minerals in the summertime, they will likely be even shorter in the wintertime. Here are some basic things to keep in mind.

Younger soils are usually more fertile and fertile soils have higher mineral levels than do eroded or highly weathered soils. The more recently your land was glaciated, the more fertile it will likely be. If your land floods regularly, but does not erode, the more fertile it will be. The longer its history of crop farming, the less fertile it will be. If it has always been grassland, the more fertile it will be due to higher organic matter and less erosion over the years. The more fertile your farm or ranch, the less need for mineral supplementation.

The more diversity in the plant community in your pastures, the broader is the range and levels of minerals in the forage. Differences in plant requirements and rooting depths generate different mineral profiles in every plant. The wider the range in ecological functional groups in the pasture, the broader is the available mineral range.

As already mentioned, crop farming lowers the mineral content of soils over time, particularly micronutrients that have not historically been replaced through fertilization. A history of manure or litter application from external sources usually helps build up macro and micronutrient levels. High organic matter content soils usually carry more available minerals than low OM soils.

The bioavailability of minerals to grazing livestock usually declines with increasing plant maturity. If you are stockpiling full season growth for winter use, you are more likely to need mineral supplementation than if only later season growth is stockpiled. In rangeland situations it is almost inevitable that livestock will be grazing full season growth so mineral supplementation is usually required. Crop residues are typically low in available minerals while winter annuals are frequently high— another reason to overseed annuals onto crop residue fields.

Higher performance goals mean the animals need more minerals. You need to decide whether the cost of supplying supplemental minerals is cost effective within the context of your performance goals. Sometimes accepting reduced performance is the more cost effective option.

My general recommendations regarding mineral supplementation are: observe your animals; be aware of chronic mineral deficiencies in your area; use soil, forage, or hair tests for diagnosis; and work with your local veterinarian or nutritionist on identifying problem areas.

Natural protein supplements

The first goal of protein supplementation is providing nitrogen to the rumen microbes so they can break down fiber and extract energy otherwise unavailable to the animal. Because we are really looking for N for the bugs, both natural protein and N in the form of urea can be effective. A second goal of supplementation is using a bypass protein to directly provide protein to the animal with a protein source not fully degraded by the rumen microbes.

Mature animals at maintenance have relatively low protein requirements (6-8%) compared to young growing stock

(10-14%) or lactating mature animals (9-15%). One of the primary objectives of calving in sync with nature is to bring winter protein requirements closer to what is available from the available forage.

As an individual nutrient, protein is more expensive per pound than energy, but a protein supplement has the capability of also increasing energy extracted from fiber so money spent on protein supplementation is, in essence, also buying an energy supplement with the same money. Energy supplements, on the other hand, do nothing to enhance protein digestibility. Most energy supplements also contain some protein, but there is not a synergistic response to get additional protein out of the base diet.

Important principles to keep in mind when it comes to protein supplementation are: 1) it doesn't have to happen every day, 2) always buy protein supplements based on cost per pound of protein not price per ton, 3) higher % protein supplements are usually cheaper per pound of protein, 4) convenience comes at a cost.

Rumen digestion is more of a batch process than continuous flow through. This allows protein to be pulsed into the rumen periodically with just as good an effect as having it continuously being made available. Numerous research studies have shown most classes of ruminant livestock perform equally well with protein supplements fed every three days compared to every day. This can save significant labor cost for delivering the supplement.

It pays to know the protein content of whatever supplement you're considering and then compare sources on the basis of cost per pound of protein.

Here are some examples. Alfalfa hay at 18% protein contains 360 lbs of protein per ton. If the hay costs $120/ton, the cost per lb of protein is 33.3 cents/lb. Corn gluten feed containing 22% protein contains 440 lbs of protein. At $180/ton, the protein cost is 40.9 cents/lb. Soybean meal at 48% protein contains 960 lbs of protein and costing $300/ton has a per pound cost of 31.2 cents/lb. So, in absolute terms, soybean meal is the best buy.

The next question is, what else are you getting with the protein? Alfalfa hay brings lower energy content than either corn gluten or soybean meal. Corn gluten feed delivers the most additional energy of these three choices. For dry cows needing a rumen supplement of coarse fiber, the soybean meal would probably be the best choice. For young growing stock, where additional energy might be very beneficial to increasing rate of gain to meet a specific target weight by some date, the corn gluten feed would be a better choice.

Another consideration is delivering the supplement to the stock. If you are already equipped for feeding hay, but not a meal, the alfalfa hay might be the right choice for you. However, anytime you supplement with hay, there will be more bulk material to transport and for the stock to consume. If your goal is to deliver one pound of protein supplement per day, which would be three pounds fed every third day, here is what you would need to haul and feed each time.

Alfalfa hay at 18% protein would require 5 1/2 lbs/day or 16 1/2 lbs every third day, not allowing for any waste. If the cattle ate all this hay when you put it in front of them, they might eat little other forage that day. They might also think that was pretty easy grub and stand around the next day waiting for more. By the time they realize you're not coming, they may already be losing a little condition. Supplementing with hay is a touchy thing when it comes to animal behavior and response.

Corn gluten feed at 22% protein requires 4 1/2 lbs/day or 13 1/2 lbs every three days. It's a little less weight and a lot less bulk to haul around. Cattle can readily consume this much CGF and go out and graze later in the day. You obviously need something other than a hay spear to feed a concentrate supplement.

The soybean meal at 48% requires just over 2 lbs/day or about 6 1/2 lbs every third day— obviously a lot less stuff to haul around and less bulk going into the rumen to inhibit grazing. There are many anecdotal reports of cattle going immediately out and grazing more aggressively after being fed soybean meal. Stimulation of microbial activity in the rumen could be a logical explanation of such behavior.

Another common misconception regarding feeding concentrate supplements is the need for feed bunks. With daily strip grazing of stockpiled or swathed forage, the need for moving bunks would be a hassle. Just the cost for bunks to supplement several hundred or thousands of head of cattle would be cost prohibitive. We, and many of our clients, have fed supplements directly on the ground.

There are at least two ways to do this effectively with minimal wastage.

One is to feed along an electric fence. This works very well with daily strip grazing if you use your lead fence each day. It puts feed on clean ground every day and has the stock line up just like being at a feed bunk. Some graziers use this as part of their preconditioning program to meet the "bunk-broke" weaning criteria. If the stock were moved to a new strip in the morning, I prefer to feed later in the afternoon so it doesn't interfere with their peak grazing time and doesn't create much forage waste along the fence where they are being fed.

Another way to feed on the ground with minimal waste is a dump feeder. This is a feed delivery system with a bin usually mounted on a flatbed pickup or 3-point hitch. As you drive along it dumps a pile of feed every so often. You can set the amount of feed in each dump and the frequency of dumps based on animal numbers and how well they clean things up. They can also be set for continuous flow so the same unit can be used for either fence-line feeding of dump feeding.

Pelleted feeds are usually cleaned up a little better off the ground compared to ground feeds or fine meals like SBM (soybean meal) or dried distillers grain. At FSRC we evaluated feeding waste with different products and amounts fed on the ground. In general, the amount of waste was surprisingly low. As you might expect, there was a tendency towards higher waste as the amount of feed delivered at any one time in-creased. When the supplement level was less at 1/2 % of body weight, the waste was usually less than 5% for either pelleted or meal supplements. At feeding levels above 1% of body weight waste approached 10% with meals and were intermedi-ate for pelleted feed.

Bypass protein are proteins not degraded in the rumen and passing directly to the small intestine where the animals can utilize them directly. Almost all forage and concentrate protein sources contain some level of bypass protein. Forages high in tannins tend to have higher bypass protein levels.

Studies with mature animals and stockers over 800 lbs have shown little or no benefit to feeding bypass protein. The greatest response comes with beef and dairy calves in the 200 to 500 lb range or with very young lambs. For the most part, I don't believe bypass protein needs to be a concern unless you are overwintering very young stock.

Using non-protein N as a supplement

The rest of the world uses NPN (non-protein nitrogen) as a supplement for ruminants far more than do USA farmers and ranchers. Part of the reason is NPN is the lowest cost option for supplementing N to the rumen and most of the rest of the world historically worked with lower prices than the USA. When I visited Queensland, Australia, several years ago, almost all of the stations we visited used urea supplied through their stock water systems as their only protein supplement. One or two were still feeding urea as a dry supplement and, while we were there, we got to see twenty-some replacement heifers fall over dead from urea poisoning.

What a nice segue for saying using urea as your only protein supplement requires timely and consistent feeding management. The problem that occurred on the Australian station was they had run out of supplement for a couple of days and the delivery truck brought them a more concentrated urea supplement than the stock were used to consuming. Knowing it was a richer formulation, they cut it with alfalfa meal and salt to dilute the urea. The problem was binge feeding because the heifers had been out of supplement, the alfalfa meal was really tasty, and using a product the manager was not really familiar with. The result was a wreck.

Urea is low cost and works well with low-protein dormant grass to create a balanced diet. Just make sure you have the right formulation and keep it in front of the stock at all times. Delivering urea through the water is a very safe and effective way of using urea if there is no other available water for the stock. If all their water consumption is coming from a tank, the daily water consumption is pretty predictable and reliable, so as long as your injection system is metered properly, the livestock should be getting a safe and consistent level of NPN.

Supplementing energy

As already mentioned supplementing energy usually becomes substitution. Energy substitution becomes very expensive because almost all forms of harvested energy are more expensive than the energy contained in standing forage, because it has to be delivered every day, and it tends to be bulky unless fed as grain.

The only time I really view giving additional energy as a supplement rather than substitution is when it is fed in limited amounts adjusted to changing weather conditions. Bos taurus (British and Continental breeds) cattle are very tolerant of cold conditions as long as they have a dry hair coat. The lower end of the thermal neutral zone, the temperature range at which an animal does not feel either hot or cold, is surprisingly low. About 15-18 degrees F is the generally accepted value although more recent literature suggests it is even lower in many instances.

With a wet hair coat, however, the lower comfort temperature is up around 40 degrees F. Windchill is another important consideration. Bos indicus (Zebu type) cattle have a

much higher thermal neutral zone and are much more susceptible to cold stress.

When cold conditions become extreme, supplementing
energy may be a cost effective and wise choice. Routine supplementation of energy just because you feel sorry for the cows
being outside when you feel cold is never cost effective nor
wise.

Energy can be supplemented by increasing the daily
grazing allotment or moving from stockpiled feed to a higher
quality winter annual forage. Supplementing energy does not
necessarily mean feeding grain or hay. Just watching animal
behavior and monitoring BCS is the best way to determine
whether or not your stock need energy supplementation.

If cattle spend only the expected time grazing (7-10
hours daily) and appear to be contented, they are probably
doing fine. If they are spending an inordinate amount of time
wandering around the pasture, going back and picking at areas
already grazed, and appearing generally discontent, they are
probably energy deficient. Adjust your management first by
allocating more stockpiled forage or changing which type of
feed you're providing.

If you do need to supplement energy, most of the same
guidelines presented for choosing protein supplements apply:
cost per unit of energy, ease of delivery, and the cost of convenience. Grain concentrates are usually the lowest cost per
calorie, unless you are in one of those areas where hay can be
purchased for less than the cost of production. All concentrate
supplements can be fed directly on the ground with appropriate
management. While lick tubs are convenient, they are usually
the highest cost source of both protein and energy supplements.

Using the fat bank

Every fat cow carries an energy reservoir on her back. It
pays to use that fat bank as it is the lowest cost energy supplement available, as long as it was created with low-cost forage
during the growing season. Estimates of the energy value of

one BCS range from about one-quarter to one-half ton of hay.

A cow calving on spring or summer grass can afford to drop down into the BCS 4 range. As long as she is regaining weight at the start of the breeding season, she will breed about as well as if she had been maintained at BCS 5. A cow going into winter with BCS of 7 is carrying from 3/4 to 1 ton of hay on her back. You just as well let her use that feed during feed stress times in the winter. Avoid using it too soon or too rapidly because it is hard to recover on just stockpiled feed.

I have emphasized the need to maintain an ongoing pasture inventory for effective year-around grazing. It is also necessary to continuously monitor stock body condition and track the trends and changes so you can make management adjustments as needed.

Chapter 26
What to do when it doesn't rain

To graze in the wintertime you must be able to grow enough forage during the growing season to carry you through the dormant season. There will be times when it doesn't rain when you expect it to and you can't grow much forage without water. Forward planning has been a recurring theme throughout this book and having a drought contingency plan is an essential part of being prepared.

Here are seven key components to a drought plan for a year-around grazing operation:

1. Know your carrying capacity with average precipitation. The best way to know your carrying capacity is through good records for your farm or ranch. The more years you keep those records, the more valuable they become as decision support information. If you are new to the property and have no grazing records, using the USDA Soil Survey to generate a predicted carrying capacity based on soil type and productivity estimates is a good starting point.

As a general guideline, we rarely try to operate at maximum carrying capacity unless you are in a strictly yearling operation that can be rapidly liquidated in the face of drought. How close you decide to operate to your potential carrying capacity depends on the frequency at which drought occurs in your neighborhood and your level of risk aversion.

In areas with reliable rainfall or very good water rights,

you may choose to maintain over 90% of your potential. That leaves very little buffer for poor pasture growth so you need to be on top of any changes before they even occur. Be prepared to destock early.

If every one-in-three years your area faces significant drought, you will probably want to take a more conservative stocking stance—maybe no more than 50-60% of your annual stocking rate for your centerpiece enterprise and 40 to 50% in flexible stocking that can be added to or liquidated easily.

2. Maintain an ongoing pasture inventory. Keeping up to date with bi-weekly pasture inventories lets you know what you have as available forage and helps predict what you will have in the near future. The inventory will help you track changes in feed supply that may be the first indicators of an upcoming drought.

Information is power and nowhere is this more true than planning for a winter following a summer of drought. Knowing at the onset of winter exactly what you have available for feed at the end of the growing season is critical for surviving the winter without a financial wreck. Keeping the pasture inventory current lets you know well in advance what options you need to be considering. The emphasis is again on destocking early.

3. Know the probability of less-than-average precipitation. Using historical weather records for your location can help you figure out whether that severe drought is a 1-in-3 probability or a 1-in-10 likelihood. Your stocking management will be different depending on the relative probability of having a drought. In Missouri we expected to have 25% less than typical stockpile one year in four. One year in seven we could expect to have less than 50% of normal.

Long term weather records for most locations are easily accessible through the Internet. You might need to rearrange them to make them usable for your particular questions and situation, but that can easily be done with spreadsheets. I like to look at years or even individual months and compare precipitation as a percentage of the long term average. You can then sort by percentile ranges to see how many years have <90%, <80%, <70%, and so on, of average precipitation. If you find 3 out of 60 years had less than 50% normal precipitation, that is a 1-in-20 year or a 5% probability of having a year like that. Not statistically perfect, but an effective way of looking at the weather.

4. Have more than one livestock enterprise. I always say marry your spouse, not your livestock. It is economically disastrous to try to feed your way through a drought while maintaining the herd you would carry in a normal year. It typically takes five years of profit to pay for feeding through one drought year. Plan to have at least one seasonal enterprise that is very easy to liquidate such as stockers or cull cows.

The key factor in making flexible livestock enterprises work effectively is avoiding the buy high-sell low wreck. One of the ways to avoid this pitfall is by dealing in ugly cattle. They can usually be bought cheaper and upgraded to a better status, thus bringing a higher price when you sell them. This does not mean buying sick and crippled cattle. It means buying off-color cattle, out-of-season classes, and anything else that brings a discount for otherwise healthy cattle. If you buy only the best looking cattle, they can either stay good looking or go downhill. They cannot be improved so you are locked into the pay scale for that particular class of animal.

Custom grazing can also work if you build flexibility into the contract. Sometimes, though, it seems harder to get out of a custom grazing deal than to liquidate your own livestock.

5. Manage for several different forage options. Having just one pasture option for winter grazing is much more likely to crash in a drought than having two or more options. Stockpile some perennial pasture in case there isn't enough moisture to get your winter annuals pasture going. Use both warm-season and cool-season forages if you live where both are adapted.

Warm-season species use much less water to produce a ton of feed than do cool-season forages. If you live in an area prone to drought, you can raise either warm-season perennials and annuals to take advantage of their inherent water use efficiency. Warm-season annuals are more drought sensitive than are perennials so you might have to plan to raise the annuals early in the season and then swath them for later use.

Renting drought stricken disaster crops from a neigh- boring crop farmer is a strategy many Midwest and Plains graziers have found useful. Crop insurance may be a deterrent here as many policies prevent the crop farmer from marketing the crop in any way that could generate additional return above the insurance payment. If you have a good relationship with your neighbors, you might be able to trade the grazing opportu- nity for other considerations.

6. Begin destocking before it becomes a wreck. Most farmers and ranchers hold on to their livestock too long before selling or shipping stock away from their drought stricken farm. I think there are two reasons for this. One is they don't realize how bad it is soon enough because they don't maintain a pasture inventory. You don't run out of pasture overnight. It has usually been coming for several weeks and we either just didn't notice or didn't pay attention to the signs. The second reason is the eternal optimism of farmers and ranchers. It just might rain next week.

7. Monitor pasture. Monitor livestock. Monitor mar- kets. Monitor other opportunities. Wrecks usually occur be- cause someone failed to have a contingency plan. Have a plan with trigger dates and conditions and pull the trigger when the time comes. Have an alternative feed plan in place. Maintaining a hay reserve will not automatically condemn you to Bad Grazier's Hell. The questions are: 1) how much should you keep on hand, 2) where should it come from, and 3) where will you get more if you need it?

I don't know who first said it, but the best drought planning strategy I ever heard was, "Accept you live in a state of continuous drought interspersed with occasional wet spells." Plan accordingly.

Chapter 27
Dealing with snow, ice, wind, cold, mud and whatever else Mother Nature throws at you

Snow

The fear of deep snow stops more winter grazing than deep snow ever has.

I remember back in Missouri, when we first started trying to push the envelope of year-around grazing, local cattlemen telling me that wouldn't work here because the snow would get too deep and the winter was just too long. When I kept asking about specific conditions, it all came back to the winter of 1977-78. That was the winter everyone around there remembered as **The Winter**.

Talk to the older guys and they could remember the winter of 1948 as being really bad. Even in Idaho, the old timers talk about the winter of 1948. One older lady from the Lemhi Valley told us about coming down to Salmon on Christmas Eve for a brief holiday visit from their ranch up on Hayden Creek and not being able to get back to the ranch again until May 9 after a big Christmas Day blizzard. It's something you remember for your entire life and it shapes your picture of winter until it becomes your paradigm.

And then we have paradigm paralysis. We come to believe something so strongly, we believe it to be true no matter the evidence to the contrary. In 23 winters in Missouri I never saw snow deep enough to stop a cow from grazing, as long as she knew that was what she was supposed to do and we allowed her to do her job. I saw ice stop grazing a number of times, but never snow.

When we moved to Idaho, we heard the same thing. There's too much snow here to graze in the winter. So far we haven't seen it in our valley in the six winters we have been here. There are a few ranchers here who have been around a lot of years and who understand what a cow is supposed to do. I've asked them how often the snow actually gets bad enough to stop grazing and their answer is not very often. I asked when was the last time it occurred, the answer was winter of 1977-78. That must have been a bad one just about everywhere.

This is not to say there aren't places where the snow gets too deep to graze. It happens very regularly in the Great Lakes region, in different locations around the West, and from time to time almost anywhere, even in Georgia, but it doesn't last very long there.

I get asked at a lot of conferences how much snow will a cow graze through. I answer with a story from a conference road trip I took across the prairies of Canada in a snowy mid-December back in the 1990s. The theme of all the conferences was extending the grazing season. I told the Canadian audience in Brandon, Manitoba, that our Missouri cows grazed through a foot of snow with no problem. That was about all we ever got out on the level. Drifts might be three or four feet, but not on the level. The Manitobans responded by saying their cows would graze through two feet of snow. Well, it's Canada and you would expect them to do a little more with snow. Especially when you see all those 1600-lb cows they have up there.

Our next stop was in Saskatoon, Saskatchewan. When I told the audience Missouri cows grazed through a foot of snow and Manitoba cows would graze through two feet of snow, not to be outdone, they told me their cows grazed through three feet of snow. This was shortly after swath grazing had become a common practice out on the prairies and it made perfect sense that a cow could easily find a two-feet-thick windrow under three feet of snow because it was only a foot down to the top of the windrow. From Saskatoon we made our way towards Rocky Mountain House, Alberta, where the snow was deep and thick. I gave them the Missouri, Manitoba, and Saskatchewan report on grazing through snow and one Alberta rancher put his hand up and said his cows were grazing through four feet of snow, eh. He provided no hay and no water to the cows. They grazed stockpiled meadows and ate snow. His neighbors testified it was true, but the cows had bloody, raw muzzles from breaking through crusted snow and sagebrush. The bottom line is a cow can do whatever she gets used to doing.

In a discussion with Kit Pharo and others regarding winter grazing, Kit made a very good point. He said push a cow to her limit but don't be stupid. That is pretty good advice.

What are some of the things you can do to make snow grazing easier on the livestock?

The first thing is to make sure there is plenty of forage under the snow. It is much easier for a cow to work through 15 inches of snow if the forage is ten inches tall than if it is three inches of bluegrass and white clover. That's your job. Make sure there is plenty to eat and she can take care of harvesting it. If you have some fields with short forage cover, use them at the outset of the fall-winter grazing season. Save the best for later in winter.

If you routinely expect two to three feet of snow, use swath grazing rather than standing stockpile.

Just like the ski slopes crave powder, winter graziers also crave light, fluffy snow. Heavy wet snow is much harder on stockpile yield and quality. Ideally, for stockpile grazing, you would like it to get cold and stay cold. Freezing and thawing causes much more forage deterioration compared to pasture just setting out there in the cold.

Freezing and thawing also leads to crusting of the snow making it much more difficult for stock to graze. If cows can get their heads under the crust they can lift and break the snow away to graze, however, they don't paw and break snow like wild game and horses. One ranch in Montana found the most cost effective way for them to keep stockpiled forage available for their cow herd was to run horses with them. A ratio of one horse per ten cows was adequate to create enough breakage in the crust for the cows to graze.

Others have run a tractor back and forth to break the snow. To do larger areas, a field cultivator can be used. These solutions put you back into the iron and oil dependency. Custom grazing horses never looked so good.

Ice

Ice is a different story. Even horses and elk can have a hard time dealing with an inch or so of ice. The only good thing about ice storms is the ice generally doesn't stick around very long and once it is gone, the stock can get back to grazing. Of course, I remember the winter when ice stayed for six to eight weeks in much of southern Missouri. Or so I was told.

In north Missouri we rarely had ice last longer then a few days. Yes, there were still heavy coatings of ice in some places, but the cattle were breaking through it and grazing. In one of my last years at FSRC, we had fall-calving cows grazing either stockpiled fescue-clover, cereal rye+annual ryegrass, or being fed hay. An ice storm hit on January 31st and it was nasty. We took little square bales out to feed the pairs on the morning of the storm and the following day because the ice was still looking pretty solid. The third morning we took bales out to the pasture, found the cows and calves all grazing even though the temperatures hadn't gotten anywhere near freezing. We took the hay back to the barn and let the cows do their job. Those two days were the only hay the cows on stockpile received that winter. We were out of winter annual forage by mid-February and those cows were fed hay the remainder of the winter.

Trucks, tractors, harrows, and field cultivators have all been used to try to break up ice cover with varying degrees of success. The biggest problem with this approach is shattering the ice-covered forage and destroying your forage supply. I have observed the higher the stock density, the more effectively livestock can deal with ice and the less time you may need to provide extra feed to them.

Wind

Wind creates windchill and decreasing windchill raises the energy requirement of all livestock. If you're in an area that only occasionally experiences gusty winds, the livestock can largely deal with it. That is why they have hides and hair or wool. Recent research from Montana State found wind to largely be a non-factor in animal performance or feed requirements for cattle on stockpiled foothills range. Out on the Plains it can be a different story.

Steady winds that routinely drive wind chills to the -20 to -80 degrees F level are hard on all classes of livestock and you need to plan to deal with it. This may mean nothing more than making sure their winter pastures have access to coulees, creek channels, or natural windbreaks. If you are grazing open crop fields you may need to construct windbreaks. Permanent windbreaks are expensive, but so are dead cows.

Severe windchills require additional energy intake so extra feeding also needs to take place. According to Kansas State research, energy requirement goes up about 0.9% for each degree of windchill below the thermal neutral zone. If we take the threshold temperature to be +15 F and the windchill is -45 F, the increased energy requirement for that day is about 54%. That is a lot more pasture, hay, or concentrate. When it gets this severe, feeding a concentrate may be the most convenient and cost effective strategy.

Or maybe the most cost effective strategy is to just be in the summer grazing business and don't keep any cows around in the winter time. Think about it.

Mud

Back in Missouri, by far our greatest challenge in winter

grazing was mud. Most of our winter was spent fluctuating around the freezing mark. It was just as likely to be 40 degrees F as 20 degrees F in January. The product was the opportunity to have soft ground and slick clay mud any day of the year. The mud usually wasn't a problem for grazing except in our lowest fields. Our solution, don't use those fields for winter grazing.

We also used to have problems on winter annual pastures that had been seeded on a prepared seed bed. Our solution for that problem was switching to no-till seeding all winter annual pastures. Very rarely did we ever have a problem on winter annual pastures after that switch. If we were in a field that was getting soft and there was still forage left in the paddock, we just moved off and came back to it later after it had dried out.

Another strategy to use for winter annual pastures is only let the stock out on them for a couple of hours in the morning and in the afternoon. If you are using the annual pasture to supplement stockpiled warm-season pastures like bermudagrass, the stock probably only need one session of grazing on the annual pastures every other day. This works particularly well if you have frozen ground in the morning that thaws by afternoon. Get them on the annual pasture at first light and move them off before the mid-morning thaw.

There is a huge difference between tall fescue dominant pastures and those consisting of mostly bunchgrasses like orchardgrass or timothy in terms of how they stand up to winter grazing in wet conditions. Plan to use fields dominated by bunchgrasses when it is drier or the ground solidly frozen while saving the tall fescue dominant areas for the wettest time. Even if a tall fescue field looks like it's taking a beating, it will usually recover in spring.

We've already talked about moving livestock quickly when it is muddy. That can't be emphasized enough. Think of it like this. It isn't the first time a hoof hits the ground that does the damage, or even the second time. It's the 100th and the 1000th hoof time hoofs hit the ground that does the damage.

The longer you leave cattle on the same area, the greater the damage, and the longer it will take that pasture to recover in the spring or following summer.

Mud was much more of a problem while feeding hay. I don't think I fully realized how nasty a problem it was until I moved to the high desert. I'm happy to say we never worry about mud anymore.

Chapter 28
Coping with peer pressure and ridicule

Hopefully by the time you have read this entire book and reached this final chapter, you're feeling highly motivated to kick the hay habit and get on with your shift to year-around grazing. The case presented for grazing as a more sustainable production strategy was sound. The economic ramifications are clear. I have shown a lot of possible options and strategies from which you should be able to pull some useful ideas and tools. Now comes the toughest part of all: actually doing it.

I have learned over the years the greatest challenge to making change in agriculture is not whether or not it makes biological or economic sense. It is what will my family and neighbors think of me and what I am doing?

Bunch quitters are never very popular. They get ridden harder and corralled tighter. Every time one comes to the chute, we beat it a little harder and give it an extra jab with the hotshot.

That is what you need to expect and be prepared to endure.

If you grew up in farming or ranching, the toughest sell is likely to be your father. He's the one who has been doing it the old way and you are essentially telling him what he has been doing is wrong. The best strategy for this imminent confrontation is discussing with him what made economic sense in 1972 doesn't make economic sense today. It's not because Dad did it wrong, it's simply a matter of the world all around us has changed. In all likelihood your father had the same conflict with his father because what made sense in 1946 didn't make sense in 1972. Your dad probably got some things changed, maybe not everything he wanted to and maybe not as fast as he wanted, but he made some progress.

There was a song by Cat Stevens from way back in the late 1960s called "Father and Son." When you're ready to smash something, listen to this song.

Choose your battles wisely and stay calm.

Farming or ranching with your siblings can be about as challenging. Every childhood argument you ever had, no matter how petty, will likely resurface as you make your case for changing the way you do business. Depending on where you are in the family hierarchy will make a difference on how seriously you're taken. Every family dynamic is a little different so I can't say whether being the oldest sibling or the youngest sibling is advantageous.

One of my older brothers was pretty bad about always telling me how I should do things. I remember working on some piece of equipment when we were both in our late teens and he really aggravated me by telling how to do something. I picked up a rather large pipe wrench, grabbed him by the shoulder, and was ready to whack him. He just looked me right in the eye and said, "Grow up, kid." I have forgiven him.

Choose your battleswisely and stay calm.

For the first few years your neighbors will only see the things that went wrong and they will be sure to point them out to you. Just remember your neighbors don't do your work for you and they don't pay your bills. After you get into it for a few years and start having some successes, they probably won't notice. If you get written up in the local paper or a major farm magazine, they will discuss among themselves everything that was left out of the article, not what was in it. Many years down the road, you may see them doing exactly what you were doing ten years ago and talking about what a great system they have developed. Don't expect to get much credit.

Love thy neighbor as thyself.

There are people out there who have already accomplished their goal of year-around grazing. They have already endured what you will endure with family and friends. Seek them out. Learn from them. There is no need for you to make every possible mistake. You will experience more mistakes than you would like, but hopefully you will be wiser for it.

Success is an attitude.

Good luck and good grazing.

Acknowledgements

My first thanks are to Allan and Carolyn Nation for their enduring patience with me for getting this project completed. It has been over five years from the time I began the first chapter until the final chapter of this book was written.

I also want to give my respects to the late Bob Evans of Gallipolis, Ohio, who kept asking me why I didn't talk about year-around grazing more. Better known for his sausage and restaurants, Bob Evans was a tireless advocate for the family farm and sustainable agriculture. Mr. Evans provided a lot of inspiration in my pursuit of year-around grazing.

There were many other farmers and ranchers all across the country who also showed me year-around grazing could be a reality across most of the USA. Too many to name here, I give you my thanks and my admiration for what you have accomplished.

Index

More from Green Park Press

AL'S OBS, 20 Questions & Their Answers, by Allan Nation. 218 pages. **$22.00***

COMEBACK FARMS, Rejuvenating soils, pastures and profits with livestock grazing management, by Greg Judy. 280 pages. **$29.00***

GRASSFED TO FINISH, A production guide to Gourmet Grass-finished Beef, by Allan Nation. 304 pages. **$33.00***

KICK THE HAY HABIT, A practical guide to year-around grazing, by Jim Gerrish. 224 pages. **$27.00*** or order Audio version - 6 CDs with charts and figures **$43.00**

KNOWLEDGE RICH RANCHING, by Allan Nation. 336 pages. **$32.00***

LAND, LIVESTOCK & LIFE, A grazier's guide to finance, by Allan Nation. 224 pages. **$25.00***

MANAGEMENT-INTENSIVE GRAZING, The Grass-roots of Grass Farming, by Jim Gerrish. 320 pages. **$31.00***

MARKETING GRASSFED PRODUCTS PROFITABLY, by Carolyn Nation. 368 pages. **$28.50**

NO RISK RANCHING, Custom Grazing on Leased Land, by Greg Judy. 240 pages. **$28.00***

PADDOCK SHIFT, Revised Edition Drawn from Al's Obs, Changing Views on Grassland Farming, by Allan Nation. 176 pages. **$20.00***

PA$TURE PROFIT$ WITH STOCKER CATTLE, by Allan Nation. 192 pages **$24.95*** or Abridged audio 6 CDs. **$40.00.**

QUALTIY PASTURE, How to create it, manage it, and profit from it, by Allan Nation. 288 pages. **$32.50***

THE MOVING FEAST, A cultural history of the heritage foods of Southeast Mississippi, by Allan Nation. 140 pages. **$20.00***

THE USE OF STORED FORAGES WITH STOCKER AND GRASS-FINISHED CATTLE, by Anibal Pordomingo. **58 pages. $18.00***

* All books softcover. Prices do not include shipping & handling

**To order call 1-800-748-9808
or visit www.stockmangrassfarmer.com**

Questions
about grazing ???????
Answers *Free!*

While supplies last, you can receive a Sample issue designed
to answer many of your questions. Topics include:
* Five Things for Farming Success
* New Composite Breeds for Grassfed Industry
* Year Around Grazing with Stockpiled Perennials &
      Annuals
* The Right Grazing Density
* Fresh Water Is the Primary Soil Nutrient
* Chopping Hay Doubles Its Voluntary Intake
* Raising Healthy Sheep
* Rare Breeds That Live Outdoors in Winter
* And more

Green Park Press books and the Stockman Grass Farmer
magazine are devoted solely to the art and science of turn-
ing pastureland into profits through the use of animals as
nature's harvesters. To order a free sample copy of the
magazine or to purchase other Green Park Press titles:

P.O. Box 2300, Ridgeland, MS 39158-2300
1-800-748-9808/601-853-1861

Visit our website: www.stockmangrassfarmer.com
E-mail: sgfsample@aol.com

Name _____

Address _____

City _____

State/Province_____Zip/Postal Code _____

Phone _____

Quantity	Title	Price Each	Sub Total
____	**20 Questions** (weight 1 lb)	**$22.00**	_____
____	**Comeback Farms** (weight 1 lb)	**$29.00**	_____
____	**Grassfed to Finish** (weight 1 lb)	**$33.00**	_____
____	**Kick the Hay Habit** (weight 1 lb)	**$27.00**	_____
____	**Kick the Hay Habit Audio - 6 CDs**	**$43.00**	_____
____	**Knowledge Rich Ranching** (wt 1½ lb)	**$32.00**	_____
____	**Land, Livestock & Life** (weight 1 lb)	**$25.00**	_____
____	**Management-intensive Grazing** (wt 1 lb)	**$31.00**	_____
____	**Marketing Grassfed Products Profitably** (1½)	**$28.50**	_____
____	**No Risk Ranching** (weight 1 lb)	**$28.00**	_____
____	**Paddock Shift** (weight 1 lb)	**$20.00**	_____
____	**Pa$ture Profit$ with Stocker Cattle** (1 lb)	**$24.95**	_____
____	**Pa$ture Profit$ abridged Audio -- 6 CDs**	**$40.00**	_____
____	**Quality Pasture** (weight 1 lb)	**$32.50**	_____
____	**The Moving Feast** (weight 1 lb)	**$20.00**	_____
____	**The Use of Stored Forages** (weight 1/2 lb)	**$18.00**	_____
____	Free Sample Copy ***Stockman Grass Farmer*** magazine		_____

Sub Total _____

Shipping	Amount	
Under 2 lbs	$5.60	
2-3 lbs	$7.00	Canada
3-4 lbs	$8.00	1 book $13.00
4-5 lbs	$9.60	2 books $20.00
5-6 lbs	$11.50	3 to 4 books $25.00
6-8 lbs	$15.25	
8-10 lbs	$18.50	

Mississippi residents add 7% Sales Tax _____

Postage & handling _____

TOTAL _____

Foreign Postage:
Add 40% of order

**We ship 4 lbs per package
maximum outside USA.**

www.stockmangrassfarmer.com

Please make checks payable to
**Stockman Grass Farmer
PO Box 2300
Ridgeland, MS 39158-2300**

**1-800-748-9808
or 601-853-1861
FAX 601-853-8087**